A CONVENIENT TRUCE

GARY MCCALLISTER

Dedication

In the diamond mines of friendship
I found you to share my winding road.

ACKNOWLEDGEMENTS

I do not know how to retrace the influences in my life that have led me to the conclusions of this book: a devout Mother, a strong Father, a wife who is both friend and lover who rescued me from pointlessness.

What a debt of gratitude I owe to teachers who inspired me and helped me learn to think more clearly: Miss Guadnola, Mrs. Robinson, Dr. Ferron Andersen, and Dr. Gerald Schmidt.

So many authors introduced me to the infinity of ideas. How could I ever hope to list them all?

Lastly, I am grateful for good friends and members of the church who have discussed these issues with me. Many have unknowingly mentored me in my quest to discover truth.

CONTENTS

A CONVENIENT TRUCE
A Cease Fire in the War Between Religion and Science
by
Dr. Gary Loren McCallister

PART I

Introduction - Why me? (About the author)

I am a Christian and a scientist. This book reflects my understanding about these two important subjects. Through these pages, I reconcile my faith with science.

In my academic studies, I have specialized in the life sciences. Consequently, much of this discussion will be concerned with questions about biology as that relates to faith. Biology seems to be the area of greatest question and conflict in today's world. I am unaware of any religious group seriously arguing over theories of electricity.

However, I am continually reading about the "conflict" between faith and science. The media has chosen to have "sides" and hold "debates". They know that conflict sells, and there are people who promote views for their own profit.

I don't think that way. I am not interested in arguing with my friends and family over any particular point of view. I don't want to feud with my religious friends.

I do want to share the fellowship and love of Jesus Christ with them. I don't care about winning a debate with my scientific colleagues. I want to discuss ideas and cooperate in an exciting and intellectual process with them. In either case, knowledge can make the world a better place for all.

I am frequently asked how I reconcile spirituality and my career. I listen to, and read, many distorted views and claims from both scientists and religious people. Most people get their science and religion second hand. There is great misunderstanding of both.

Many scientists ended their religious instruction and study at about the age of ten. Consequently, they may hold curious and erroneous views of religion. Religious people often do not understand how science works and are, therefore, subject to misleading information. I will examine issues on each side of this spectrum.

I am in a unique position to write this book. I am a devoted Christian with more-than-average exposure and training in theological subjects. I was raised in a fundamental Christian congregation and later joined the Church of Jesus Christ of Latter Day Saints. I have been a practicing member of this congregation for more than forty years. I have more than twenty hours of college credit in theological subjects, have taken numerous courses in the humanities, and have taught in a local seminary program for four years. In addition, I have served as a Bishop of a local congregation for several years and have been an active lay minister for over forty years. As a student of the scriptures, I hope I can bring a reasonable voice to the subject.

On the other hand, I am a scientist who has taught science courses at both the college and graduate-school level for over forty years. I have published numerous scientific papers, science education articles for the lay person, and have written an award winning science education column for our local newspaper for many years. I have been a technical writer for an engineering firm. I have served on a local school board and been involved in teaching science educators. Besides common biological courses, I have also taught robotics and computer science. I was chairperson of a science department for many years and enjoyed the practice of a scientific career. Additionally, I have been involved in regional vector-control operations and environmental concerns for almost twenty years. I hope to represent the sciences fairly.

Chapter 1 – The Order of Understanding

I don't know . . .

    I don't know what a person who "understands" looks like. I mean, how can I tell? I see students all the time who look attentive, take notes, nod, smile, and then fail exams. I thought they understood, but they apparently didn't. See, I just don't know what someone who "understands" looks like.

    I can know if someone understands something by one of two ways. Either they correctly recognize examples of the idea, or they are able to use the idea to do something.

    So how can I know if someone is a scientist, or not? What kind of scientist is he or she? How can I know if someone believes in God, is a Christian, a Jew, or a Muslim? I have to see what they do. How can I tell if someone is a careful thinker? How can I tell if someone is lying? How can I identify someone who has an agenda? I have to see what they do, and see if they can recognize correct examples from incorrect ones.

    Because so much of science consists of interwoven thought, it is difficult to know where to begin in teasing it all apart. How can we talk about neurotransmitters if we do not know some chemistry? How can we know chemistry if we do not understand some physical laws? By the same reasoning, how can I understand physics unless I know where the study came from? And how can I understand the origins of physics if I do not understand the origins of reason, logic, and materialistic thinking?

So I have organized this book around what I describe as an "order of understanding".

Sequence

It is generally assumed in the twenty-first century that religion and science are at war. This perception has been carefully fostered by people who benefit from such dichotomy. However, my hypothesis is that that is not necessarily true, nor has it ever been. Therefore -

Part II of this book addresses the origins of both religion and science and the perceived conflict between the two subjects. This will include philosophy, reason and logic. After all, most scientists receive a Doctor of philosophy degree in spite of the fact that many of them never take a philosophy course.

I will discuss the origin of the Universe, the creation of the earth, man's place within the Universe, the age of the earth, the nature of science, the idea of the soul, and the natural limits of religious and scientific methods. A chapter will be devoted to each of these issues.

However, these topics cannot be studied one at a time as total understanding necessarily crosses over disciplines. Much of the discussion involves the physical sciences of chemistry, geology, physics.

Part III of this book deals exclusively with questions having to do with life, biology, the environment, and evolution.

In Part IV I will discuss the real war that is on-going; that between religion and atheism.

To achieve a broad understanding of science, one must first understand the physical sciences. Specifically, the laws and topics that comprise physics will be addressed. These concepts then lead to a broader understanding of chemistry. Understanding both of these subjects is necessary to appreciate modern biology and geology.

Science is not taught in this sequence in the public schools of America. I am told it is in many of the industrial countries of the world. In America, the science curriculum was established in the early 1900's when biology was considered the study of nature and agriculture. Because people were surrounded by these subjects, it was thought to be the perfect starter subject for science study. Modern biology, however, depends completely on an understanding of chemistry and physics. Unfortunately, the curriculum has not changed, but I discuss physics and chemistry first, as seems logical.

Before addressing specific areas of interest shared by religion and science, I want to dispel a couple of common myths concerning their supposed conflict: the myth of the flat earth and the myth of the religiously-persecuted scientist.

Myth of the Flat Earth

It is actually Washington Irving's fault. Yes, I'm referencing the man who gave the world "Rip Van Winkle" and "The Legend of Sleepy Hollow". It appears to be mostly his fault that the world generally believes that the early Christian Church taught, and the world believed, that the earth was flat. (Irving, 1828)

10

In 1828 he published a book entitled "The Life and Voyages of Christopher Columbus". This was a fictional account but was mistaken by many as a scholarly work. In this book he sets up a fictional scene where Christopher Columbus is meeting with a commission, established by the King and Queen, to review his proposed voyage. Some of the members of the fictional commission objected to his plans on the supposed scriptural basis that the earth was not round. The scene is entirely fictitious.

In fact, the earth had been known to be round by educated persons from the 3rd century BC onward. Many Greek writers knew and recorded that a ship's body disappeared before the mast because of the curvature of the earth. Aristotle cited numerous observations verifying that the earth was spherical.

Jeffry Russell, speaking to the American Scientific Affiliation in 1997 (Russell, 1991), said that "... with extraordinarily few exceptions no educated person in the history of Western Civilization from the third century B.C. onward believed that the earth was flat". Of course, Washington Irving was not a scientist, but an author. He was among the first American authors to win acclaim in Europe, and have writing as a career. Some people mistook his fictional account as history and perpetuated that belief as a modern myth. It can even be found in textbooks.

Myth of the Persecuted Scientist

"I believe the idea that Galileo's' trial was a kind of Greek tragedy, a showdown between blind faith and enlightened reason, to be naively erroneous". So says Arthur Koestler, popular and successful author and biographer, who wrote "The Sleepwalkers, an account of Kepler, Copernicus, and Galileo". (Koestler, 1990) In his book, Koestler examines the actual circumstances of Galileo's life and his infamous trial. But his conclusion is resoundingly different from the popular perception of today.

This, primarily because in the 1930's, Bertolt Brecht wrote a play entitled "The Life of Galileo". (Brecht, 2008) The play is a black and white, highly fictional depiction of powerful religious bigotry. The lone hero is the virtuous scientist. The play was made into a film in 1975 by Joseph Losey, and this portrayal is what is now believed to be true by people today. In fact, it is believed to be only one example of repeated persecutions of scientists by the Christian church. I have had people tell me that Copernicus was persecuted by the church and burned at the stake. The truth is that Copernicus was never bothered by the church in any way.

The true story of Galileo is complex, and I recommend reading Koestler's book, (Koestler, 1990) or any of several others that explain the entire circumstance. But the story, in a nut shell, is that Galileo signed an agreement saying he would not teach the "heliocentric theory" for a variety of complex reasons. The heliocentric theory is that the earth revolves around the sun, not the sun around the earth.

He did not tell anyone about this agreement, and the agreement was not discovered until he broke his word and published a book on the subject. In fact, he tried to conceal the contract until it came to light just before his trial. He had obviously tried to keep it from being generally known.

Because of this, he was perceived as a liar, a sneak, and a man who broke his word. Then, speaking in his own defense, he also maintained that his book did not teach the heliocentric theory, when it obviously did. The court had no choice but to decide that he had broken his contract, and convict him of such.

The church should never have brought him to trial but did so for historical and social reasons. One reason being that some of his arguments were not valid, though the theory itself was. This was new science and not fully accepted. However, he was treated with respect and restraint. He, himself, exhibited poor behavior. Of course, he was right about heliocentric theory, although many of his arguments were in error. It was in this way Galileo brought about much of his own downfall.

He was never placed in a dungeon, and he was never tortured. He lived for many months in a luxurious palace and then was allowed to return to his own comfortable and spacious villa in Florence. He was allowed to travel, visit family and others, as well as to continue his scientific work on other matters. He lived comfortably and died peacefully in his own bed at the age of 77.

First battle

But the war between religion and science did have its beginning. We know who started it, and when. His name was John William Draper, a physician and scientist born in the early 1800's in

England, who later moved to Virginia with his mother. In 1874, he wrote a book entitled "History of the Conflict between Religion and Science." (Draper, 1874)

His was a sloppy application of early Darwinian thought to social and political subjects. It was also the origin of social Darwinism theories of ill repute. It is full of lies, distortions, poor conclusions, and misunderstood applications. However, the book was highly influential at the time, and the title gave birth to the thought that a conflict between science and religion existed. Today his book is mostly seen as an early, anti-religious attack published by a scientist.

Evidence of Things Not Seen

Modern man most often sees time as a progression along a horizontal line; always going forward and always progressing to a higher level. However, straight lines can decline as easily as trend upwards.

Throughout history, many people have seen time as circular. They trace time through the day and night cycle, the seasons, and even broad historical cycles with repeated trends.

The atom, as a particle, was proposed in sixth century India, and again at intervals by different cultures and philosophers for several thousand years. Presumably our modern understanding exceeds the mostly philosophical concepts of an earlier time. However, it gives one pause to read Hebrews 11:3 in the New Testament and wonder how much Paul understood about modern chemistry. Then again, we can wonder about how much modern chemists rely on their faith that this statement is true.

"Through faith we understand that the worlds were formed by the word of God, so that things which are seen were not made of things which do appear."

Part II - SCIENCE

Chapter 2 - Introduction

I don't know . . .

 I don't know how many species of solitary bees can be found on the western slope of Colorado. I don't think anyone does know. No one has made a good, comprehensive study of the insects in general in this area. I am sure the common ones are known, but there are many less-common ones in the mountains and deserts that no one has cataloged. In fact, there could be new species "out there" just waiting to be discovered.

 Speaking of new species, for the last several years, a few of us have been looking at tiny mites that live on mosquitoes and black flies. These mites are so tiny that eight or ten of them can cluster on one single mosquito. We have many specimens and are taking electron micrographs of them this year. The problem is, though we can find them readily and in pretty big numbers, no one can identify them as to species because they are immature specimens.

 By the time people finish high school, they have spent twelve years being told what humans know. It is a little understandable if they come out of school thinking they know everything. In fact, many people do pretty much assume that humans know everything. A common perception amongst people is that if you memorize enough facts and science texts, you are a scientist. If you memorize the Bible in Latin and Greek, then you must be a theologian.

Interestingly enough, many people, including scientists and theologians, don't know what we don't know. That is partly why I wrote this section about things that I know, things that humans know, and things that humans don't know; and there is so much we don't know. As a reminder of this fact, I have begun the first section of each chapter with some of the things we don't know related to that subject.

Keep in mind that there are many different ways of knowing something. Science provides one way, of course. But finding answers isn't a war. It's a quest. Other ways of knowing may, or may not, be more truthful.

Is This Important?

It is interesting to note that any discussion over religion and science is seldom about the science. It's usually about whether science harms faith, or whether religion harms science. Some people have proposed that the two are separate spheres of knowledge that should not overlap, so saying that one has nothing to do with the other. It is probably too late for that. Public education has exposed nearly everyone to science, though science is actually a relative newcomer to the world of influence. Religion has been far more influential in shaping our modern world, and it has done so for a far longer time. To say that these are separate spheres of influence and do not concern each other is not reasonable.

The resulting confusion is rampant and affects our lives whether we wish it to or not. If you want your child influenced by your personal faith, you have a stake in science education. If you think science is being used as a political foil, you have an interest in science. If you think religion has harmed people's ability to think or act freely, you have a stake in religious education. If you think religious individuals may impede scientific understanding, you should have an interest in religion.

I intend to talk about God, planets, animals, molecules, rocks, and plants all at the same time. There is no requirement for keeping them separate and there may be harm in doing so. As one famous scientist is to have said, "We don't have to believe anything that isn't true", (Eyring 1983).

People must realize that not all religion or all science is rock solid. There are squishy parts of science that we don't understand just as there are solid parts to religion which are difficult to refute. Of course, there are Christian teachings that are nearly universal and some that are accepted by only a few sects.

There are scientific principles that are useful and true. There are scientists who over-reach their data for personal gain or aggrandizement. There are religious leaders who purposely misrepresent scientific facts for their own benefits. I hope to provide a balanced view that examines both.

Why Doesn't the War Ever End?

There seems to be an increasing hunger for religious discussion in the world. People seldom ask me about my job as a scientist until they find out I am also religious. There seems to be a desire for greater understanding of our crazy, fast-paced life that has had our traditional anchors of place, family, and

18

principles supplanted with ever changing science and technology. People yearn for more peace in their lives as they watch the increasing development of technological warfare.

I think this may be why many people want to see creationism taught in the science curriculum. Political correctness has inhibited the public discussion of religion but given license to science education. Therefore, people are hungry to talk about God and see the classroom as a way into the public arena. In this book, I will talk openly about God in the same breath as I talk about animals, plants, and computer science.

Science is not an explanation of God, nor do I believe God needs science in order to exist. Here I must add that science has generally been seen as beneficial to humanity. However, knowledge and power have also been put to evil uses, and technology often creates as many problems as it solves. On the other hand, science itself has helped millions of people to live more comfortably, live longer, and to accomplish more. Much good has been accomplished in the name of science.

Peculiarly, science is silent when bad things happen to people. It isn't that scientists don't care or that science is bad. It is simply not comforting to have it explained that your child died because of toxins produced by an overwhelming growth of bacteria. It has little that is useful or comforting to say about the projectiles traveling at 2500 feet per second that were fired by the gunman in the tower as he shot seven people.

I want to go on record as saying that the war isn't between science and religion at all. I will point out the real terrorists, suicide bombers, and aggressors in this "so-called-war" in the last section of this book.

The Breadth of Discussion

I will explain certain scientific principles, but I will often quote the Bible. I will use the Bible because it is the religious reference with which I am familiar. I am certainly not qualified to speak for other religious factions. In fact, I do not pretend to speak for any Christian sect or congregation. I speak only as a lay person, who is also a scientist, trying to bridge a significant cultural gap. I will quote religious texts because there is often great misunderstanding of what is actually said in the scriptures, as well as there is often great misinterpretation. When quoting scripture, I use the King James Version of the Bible. I don't want to digress into the pros and cons of other translations or other esoteric religious arguments.

I hesitate in expressing my own religious sentiments because this is not a book about me. It is a book about trying to bring broader understanding between people of different persuasions. I hope to show that there are many possible explanations and many sources of information in both camps. Apparently, God wants us to be able to live with ambiguity.

Personally, however, I believe there is a God. I believe that Jesus lived and died for our sins. His teachings have the authority of God behind them. His vision seems to be one of love and forgiveness, not rules and details. I believe that God still reveals His will in many ways: sometimes through the inspiration of scientific discoveries, sometimes through artistic expression, and sometimes through His revealed will to individuals.   I do not intend to prove that God exists in this book. I do hope to help myself, and others, live rich, intellectual, and Christian lives.

Motives

I should point out that there are people with powerful motives who want conflict between religion and science. Some are simply ignorant. As a general rule, the scientists I know stopped attending church, reading the bible, praying, or paying the slightest bit of attention to religious subjects between the ages of ten and twenty years. These are the years when they could begin resisting parental influence. Many never had religious influence at all. So their criticisms of religion are often based on hearsay and misunderstandings. Interestingly, most people also stop drawing at about the same age, and so most people draw about like they did at ten years of age and simply say they lack talent.

Religious leaders often find themselves in the same situation. We have friends who send their sons off to a church boarding school at the age of fourteen. Many do complete public school, but they often take minimal science. And "school science" does not ever have much to do with "science". In their preparation to become theologians and ministers, they emphasize religious studies such as languages and doctrine. I have never met a minister who has conducted true scientific experiments, although I would guess there are some.

People who promote conflict may have other motives as well. If one has devoted nine years past high school to studying evolution, if one's whole identity and success depends on his work in evolution, or if one is paid to teach or do research on evolutionary topics, it is very difficult for them to be open minded about alternative hypotheses. By the same token, a minister who has spent his life learning doctrine and defending his faith, and who is paid by his congregation to defend their doctrine, is not totally free to be open-minded.

Classrooms of either of these groups of people are not going to present fair and balanced presentations. In fact, if either group tried to be fair-minded, they would probably find their credentials attacked in both camps. They would be seen by their colleagues as not being legitimate scientists or theologians.

One credential I have to write this book is that I am of retirement age, and attacks from either side on my legitimacy will have no real effect on my life.

Then there are motives such as money, power, and fame. Many militant atheists have done quite well in the last twenty years publishing books attacking religion. Many of these are scientists or intellectuals who have reputations in their fields. However, if you read their books, you will find gross errors of scholarship, illogical arguments, outrageous claims, and lack of reason.

So why are their books such money makers? Because they are controversial, and controversy sells. Then once they have a successful formula, they and the publishers have a vested interest in yet another book. These books have a powerful influence on some people and have nearly convinced many in the general population that religion is evil or delusional, at best. Yet many have done little or no real science in years.

Equally there are theologians who have become rich and powerful within their own communities, by taking hard stands against subjects they do not understand, in order to appear to be defenders of faith. Their books, sermons, revivals, and videos are every bit as biased and misleading as the scientists who have abandoned science to become propagandists.

Both sides have distorted history, and their own biases lead to argument. There are few people who are trying to see through the hype and conflict of both sides to learn the truth.

Cast of Characters

The term theologian is a category or pigeonhole. The term scientist is a pigeonhole or category. There are many types of religious people who have very little uniformity in their basic beliefs. Scientists also make up an individualistic group of people with many different understandings. It isn't possible to lump people into categories as if they possessed some kind of uniformity of thought.

However, in recent years, the war of words has escalated. There are individuals in both camps who stand out as spokesmen for their causes. Since it is my thesis that the entire war does not necessarily exist, I don't want to be caught in a battle of words between various religious denominations, scientific specialties, and outspoken individuals.

While I will occasionally quote individuals who are particularly eloquent in making a point, I do not want to be seen as attacking them. There are specific individuals who continually attack in a blatant, self-confessed effort to continue the needless war. Unfortunately, these individuals are guerrilla-warfare fighters and do not disclose their true allegiances. They melt back into the cover of innocent bystanders when they are threatened. They do not make clear that the war is not between science and religion, but is something else entirely. The only exception to this policy will be restricted to the last portion of this book. In that section I will name names and explain the true nature of the enemy.

Chapter 3 - In The Beginning

I don't know . . .

I don't understand the idea of beginning. But then I also don't understand the idea of eternity. If something is eternal, it has no beginning. But if there is a beginning, what came before the beginning must have been eternal. I get so confused.

But I think beginnings are always interesting. Sometimes they are dangerous and mysterious, like the opening murder in a mystery. Sometimes they are romantic, like how a married couple first met. Sometimes beginnings are hard work, like the birth of a child.

Perhaps it is impossible to ever know the beginning, since every beginning had a precedent cause. But I'll talk more about that later. Right now, I think it's important to understand where the various religions came from and when, and how, science was born.

Religion

Religion apparently came first. (Noss, 2011) Early man seemed to think he lived in an enchanted world. Stones, trees, and animals were inhabited by spirits. Pagan beliefs of animism, magic, fairies and goblins held sway. These spirits often seemed capricious, and the world seemed like a mysterious place. These early societies developed many practices for the purpose of trying to placate the many spirits in controlling their world. Some of what they appear to have done seems strange to us today.

A series of polytheistic religions followed. These religions claimed there were numerous, divine beings who were involved with humans in different ways. Sometimes it was a very personal involvement, and sometimes it was as the causative agent of natural disasters. The Greeks accepted at least fourteen major gods with their varying personalities and domains. Rome was a polyglot of religious beliefs, but most people recognized numerous gods and divinities. The Egyptians had as many as 2000 gods and goddesses. Some were worshipped throughout the whole country, while others had only a local following.

After those two ancient cultures, the eastern religions came into existence: Hinduism is one of the oldest. Buddhism and Confucianism came a little later, although Confucianism is not really a religion but a very ancient set of moral teachings. These religions continue today in many parts of the world. They comfort people and provide direction in people's lives. However, they have not had the global impact of the three monotheistic religions of the world.

Monotheism and Christianity

Next, historically, were the three dominant monotheistic religions of today: Judaism, Islam, and Christianity. (Noss, 2011) The difference between these three religions is interesting.

Judaism and Islam seem to be primarily religions of law. There are edicts given, sets of laws, which have the weight of divine authority. Both Jews and Muslims often argue among themselves over the interpretation of the law and how to best apply their written codes. They both appear to be based upon the practice of legal interpretations, a kind of divine jurisprudence. For Jews, the laws apply only to God's chosen people. For Islam, the law applies to everyone.

Christianity, on the other hand, is based upon personal belief, not law. There is respect for the law; but doctrine, the interpretation of the law, becomes the center of worship and is the set of beliefs about man's relationship to God. Thus Christianity is based upon theology, not law.

Neither Judaism nor Islam have theologians in the same sense as Christianity because they do not try to understand God's purpose. God has spoken and His laws are revealed. If He interferes in human affairs it is to do His will, and men cannot understand. Christians believe all have a personal relationship with God and need to understand Him personally. Christians believe every man is a theologian.

But what does a theologian do? There is some debate about the meaning of the word theology. Its origin was in the Christian tradition although the word was taken from early Greek. Because of the unique role theology plays in Christianity, some scholars restrict the use of the term to Christianity.

A working definition is that a theologian uses reason to understand the ways of God. Christianity is the only dominant religion of the world that depends upon reason and logical thought in trying to arrive at the truth about God. This does not mean that other religions do not, and cannot, use reason and logic. The others use reason and logic for different purposes, to understand the law.

Interpretation of God's message was placed in central control for many centuries. The individual relied upon Church leaders to interpret God's message and then tell them what was expected. However, in 1384, John Wycliffe finished his translation of the Bible into English. Such translations were forbidden for many years following, but a movement began advocating that common people be allowed to read the Bible in their common language. This movement spanned another two and one half centuries under the influence of such men as William Tyndale and Martin Luther, but it culminated in the publication of the King James Bible in 1611.

During these middle ages, the medieval monastery was a place to retrieve, collect, and preserve classical knowledge. Much of this knowledge had been destroyed by the barbarian invasion of the Roman Empire and Europe. But for centuries, the Monks in many Monasteries transcribed and copied earlier works.

Monasteries were self-contained units that grew their own food, made their own clothes and were pretty much self-sufficient. The Monks also had the space and time needed to try new methods of agriculture as well as research other economically important activities. Many new techniques and implementations were developed under the monastery system where trial and error was an accepted practice.

Eventually, from the Monasteries, came schools of various kinds. By the twelfth century, Universities began to be established. All early Universities were associated with the Christian church. However, they soon became independently governed. The curriculum of early Universities was, of course, religious; but it also became secular to accommodate economic and technical demands of the day (Haskins, 1957).

About the same time, at Oxford University in the 13th century, inductive and experimental methods were first used systematically to advance knowledge. Francis Bacon, "founder of the scientific method" and a devout man who wrote theological papers as well as scientific, argued that man had a divine responsibility to use his reason to establish dominion over creation. The first medical research institutions and the first astronomical observatories were all built by the Christian Churches. He argued that since Christians believe man is created in God's image, there is a spark of divine reason in humankind (Schmidt, 2001).

Science

The study of "science" came much later through intellectual and cultural development, and it is a unique cultural invention that has only occurred once in history. Historians argue about the actual beginning of science. It has roots in the 13th century, although many say it began in earnest in the 1700's.

Two schools of theological thought were operating between the twelfth and seventeenth centuries. One maintained that experimentation was the way to learn God's mind and design. The other held that scholastic debate, using the rules of deductive reasoning, was the way to discover God's will.

But note that both schools of thought sought to understand God, not deny Him. And both have contributed to our modern concepts of science. This point becomes important later in this book when we discuss the philosophy of science (Whitehead, 1953).

Though there is debate about exactly when, and precisely where it originated, scientific methodology has been applied since the middle ages, even as early as the twelfth and thirteenth centuries. Most historians describe the sixteenth and seventeenth centuries as the time of the scientific revolution. The reformation, the Bible becoming available in English, and the birth of the scientific process all developed simultaneously.

There really is no question that science arose only in Europe and only under the influence of Christianity. As Alfred North Whitehead observed in his book, Science and the Modern World, "Faith in the possibility of science . . . is an unconscious derivative from medieval theology" (Whitehead, 1953).

You see, of all the religions, only one has ever made the claim that the world is an orderly place, that the world was created with some kind of plan, that things were supposed to make sense, and that we ought to be able to understand what went on based upon reason. Only Christianity is concerned with doctrine: what is thought to be the true relationship between God and man. Man must use his reason to understand his relationship with God. There are no theologians in the other religions because man's relationship is already set through the law. The equivalent position in Judaism and Islam is to understand the law. The function of theologians, then, is to apply reason, or careful and methodical thought, to the great questions that trouble man. Why did God create the Universe? How? Why do I exist? Where do we come from and to where do we go? What is to become of us?

The following is a list of scientists who were not only Christians, but who publicly announced that their scientific enterprise was to reveal the true nature of God and his plan for the Universe. Many were also clergymen.

Copernicus	Kepler	Galileo	Brahe
Descartes	Boyle	Newton	Leibniz
Gassendi	Pascal	Mersenne	Cuvier
Harvey	Dalton	Faraday	Herschel
Joule	Lyell	Lavoisier	Priestley
Kelvin	Ohm	Ampere	Steno
Pasteur	Maxwell	Planck	Mendel

Science, which purports being the ultimate use of reason, is literally born of Christianity. Science as a methodology occurred just once in all of history. It did not occur in Greece, Rome, Iran, Israel, or China, though all may have contributed to its development. It was born in Europe from a Christian tradition of reason and thought.

This was not an accident. Only Christianity encouraged, or even allowed, questions to be asked concerning personal relationships with God that were pursued through reason. Indeed, it was expected only in Christianity.

Several scientists in the last twenty years have become militant atheists and attacked all religion in vicious terms. According to them, religion is the cause of all wars, death, famine, and ugliness in the world. They have had much success in popularizing these ideas. However, this claim displays ignorance of history as well as a lack of rationality. And rationality and reason are the crowning attributes of both science and Christianity.

The Faith of the Scientist

It is impossible to prove that the universe is rational. Belief that the universe is rational is an assumption. Modern man takes for granted the idea that the universe is governed by natural, physical laws. We assume that we can grasp these laws, that we have discovered many of them, and that we can yet discover more. We believe that we can comprehend these laws of order, and we seem to use them to our advantage. We further assume that these laws are uniform and apply everywhere in the universe. Indeed, we believe they have applied throughout all time and will never change. Further, many of these laws appear to be expressed in the language of mathematics.

These assumptions are not provable, yet they are little short of amazing. The presently accepted theory of the origin of the universe postulates a nearly infinite explosion. Why a violent, chaotic explosion would give rise to an orderly universe is not understood, and such a belief is astounding. To propose that this very explosion was caused by laws that must have existed prior to the explosion itself is a remarkable leap of faith. Yet science routinely begins with the existing order and does not assign its source.

Christianity did not invent reason or the idea that the universe is a logical and rational place. Pythagoras, Thales and other such early philosophers postulated a universe that was understandable according to natural law. Their writings had a great influence on Plato who went on to influence much later thinking. But much of their influence was lost for centuries until after the idea was resurrected by early Christianity.

Christianity has no monopoly on logic and reason. Successful scientists from many faiths, and those with no religious faith, have made significant contributions to modern science. The point I am trying to make here is only that science was born and nurtured from a specific world view introduced through Christianity.

"In the beginning was the word . . ." The term used for "word" in the ancient Greek version of the Bible was logos. Logos means more than word. In Greek it can be translated as "thought" or "rationality". Early Christians believed God created a Universe that operated according to divine laws of order and rationality. They believed that if we could understand those laws, we would know more about the nature of God.

Christianity resurrected the idea of an orderly universe because it purported that such a world had a rational cause: God. Christianity teaches that the world operates according to divine reason. Since we are created in God's image, and God is a logical, orderly, reasonable God, we must also possess these attributes. The universe is derived from God's reason, and human reason is derived from the same divine intelligence.

As discussed, under the encouragement of the monasteries, guilds, and associated schools of the 14th century, the waterwheel, windmill, chimney, eyeglasses and the mechanical clock were all invented. The first professional scientists can be traced back to the late middle ages and then again from the 16th century when the reformation occurred. With the reformation came the idea that knowledge was not the property of the ecclesiastical authorities but was something each man could decide for himself. Thus, an individual mandate for independent thought was begun.

But the universe doesn't have to be reasonable. It would be easy to imagine a cosmos where stars, billions of light years away, might have different laws from the ones that apply to the observer on earth. Early man had no trouble imagining a universe that was unpredictable from moment to moment. In fact, early man apparently assumed that very notion. He imagined a universe where things could pop into and out of existence from time to time. In fact, many fiction writers have described such worlds. There is no logical necessity for a universe that obeys the rules.

It is even odder that many of the rules governing our universe obey the laws of mathematics. Richard Feynman, the noted physicist, has said "why nature is mathematical is a mystery. . . The fact that there are rules at all is a kind of miracle" (Feynman, 1998).

Reason

Reason itself, by definition, is based on assumptions. Rational thought is a process of certifying an idea because it agrees with, or is deducible from, premises we already believe. For example, the theorems of geometry are accepted because they follow from previous definitions and accepted ideas.

Consider balancing your check book. Because you assume the information you have about your beginning balance and outstanding checks is accurate, it can be reasoned that a final balance can be made.

If one is very sure of a premise, a good conclusion drawn from a pattern of deductive thought is more likely to be accurate. But it is difficult to always be sure of beginning premises. Premises must often be derived from non-rational sources. Non-rational sources are usually referred to as "articles of faith" in religion. The belief in a uniform, rational universe is an "article of faith" in science.

There are scientists who fail to recognize that because they begin by assuming there is no God, they have no explanation for the orderly laws of the universe, laws that they assume exist. It is just as reasonable to begin with the assumption that the universe is orderly, and the cause of the orderliness is, indeed, God. Many famous scientists of the past have done exactly that.

Copernicus called astronomy, "a science more divine than human. . ."

Boyle left money in his will for a series of lectures to combat atheism.

Newton wrote commentaries on the Book of Revelations. He also said of his discoveries, "this most beautiful system of sun, planets and comets could only proceed from the counsel and dominion of an intelligent and powerful being."

Kepler wrote, "For a long time I wanted to become a theologian. Now, however, behold how through my effort God is being celebrated through astronomy."

Biologist Joshua Lederberg said in an interview with Science magazine, "What is incontrovertible is that a religious impulse guides our motive in sustaining scientific inquiry." (Tarnas, 1993)

I know some think that the scientists of an earlier day had incomplete information. Or one might assume that those scientists simply accepted the mind set of their day, and religion really had no effect on their theories and thinking.

For example, when Kepler found that the planets move in elliptical orbits, he was criticized by others because they thought God would have used perfect circles for his divine plan. But Kepler was convinced that there was an even more beautiful explanation than perfect circles and set about to try to find the explanation, not in spite of his faith, but because of his faith.

Eventually Kepler proposed his Three Laws of Planetary Motion.

1. The orbit of every planet is an ellipse with the Sun at one of the two foci.

2. A line joining a planet and the Sun sweeps out equal areas during equal intervals of time.
3. The square of the orbital period of a planet is directly proportional to the cube of the semi-major axis of its orbit.

This third law is especially amazing. How in the world, or not in the world, did he discover this relationship between the sun and the planets? Setting aside the "how", let's look at the "why".

He made these discoveries because he was convinced that God was a reasonable being and had created the universe according to understandable, mathematical laws. He reasoned it out because he was religious, not in spite of his religion.

One of the tasks of scientists is to look for pattern, order, and relationships in nature. They use many different means to measure the accuracy of their data to know whether or not they are on the right track. Often they apply something called "Occam's razor".

Occam's razor is usually attributed to the 14th-century, English logician, theologian and Franciscan friar, William of Ockham. Basically it states that if there are multiple possible explanations, the simplest explanation is usually correct. However, by simplest he means the explanation that has the fewest new assumptions; reason always relies on assumptions.

Modern scientists often express acceptance of, or admiration for, theories because they are "simple", "elegant", or even "beautiful". The religious impulse is still present in science and is expressed by beauty, simplicity and a tremendous faith that the universe is

an orderly and logical place. The source of that order
has never been completely acknowledged.

Chapter 4 - Origins of the Universe

I don't know . . .

I don't actually know when I was born. The official record says the event happened on Feb. 26, of 1945, but I don't really remember it myself. I accept the timing because that's just what I have been told.

I don't really know when my parents were married either. I wasn't born then. In fact, they had other children by the time I came along. But I have been told that their anniversary is Feb. 25. See, for much of my youth, I celebrated my birthday on some other day than the 26th because my parents were usually out of town for their anniversary on the 25th. I always joked that I messed up one single anniversary, and so they messed up years of birthdays in spite.

The point is, we humans depend a great deal upon authority figures to tell us the truth. We believe a lot of things about which we have no personal experience, knowledge, or memory.

Origins

Can we ever know the beginning? Every event would seem to have a cause. But the cause of an event has a cause, which causative-event also had a cause, and so on. This is one of the problems Thomas Aquinas wrestled with as he tried to understand the origins of the universe.

He reasoned that before today was yesterday, and before that was the day before. But there was a day-before, there was another day-before, and so on. Do the series of the days-before-yesterdays extend infinitely into the past? If so, what came before the first day before yesterday?   Since we cannot imagine that there was a day with no day-before, we assume there must have been a beginning. But what caused the beginning? And what caused the cause of the beginning?

It appears that either no one had asked this question before Thomas Aquinas, or if so, they had never recorded the answer (Aquinas, 1998). That revelation was simply that God created time when he created the material universe. And if God created time, then he exists outside of time and is eternal. This does not necessarily mean that He lives eternally, although He may, but in the sense that time does not apply to Him.

Interestingly enough, modern physics has come to a similar conclusion: that time is a property of the universe itself and came into existence at the same instant the universe came into being. In fact, time cannot exist without a beginning to measure it against. Time came into existence at the same time that the material universe came into existence. Ironically, time, space, and energy are the three subjects that have occupied most scientific thought, as well as theological thought, throughout much of history.

Many modern atheists question that, "If God created the universe, then what created God?" But they miss the point. It is true that time is composed of a series of events, but every series depends on something outside the series in order to be real. Suppose I write a novel. The events in the story form a series. Let's say the events seem to make sense and appear logical. In the story, one character causes the death of another. Does that mean that because I wrote the story, I am a murderer? No. The rules of causation are different in my story than they are for me because I am the creator of the novel. I exist outside the novel. My life is not an episode within the novel.

God is the creator of the universe. As such, He is not part of the universe, just as I am not a part of my novel. The creator is always outside the creation although intimately involved in the process itself.

Oh, who cares?

Well, actually, a lot of people care. Humans have been arguing about the origin of God, the origin of the universe, the origin of man, the origin of the species, the origins of various wars, or even the origin of a given car accident, for a very long time. Some people maintain that we can never know the beginning, and I suppose that is generally true.

Carl Sagan, an astronomer, has made the assertion that "the cosmos is all there is, or was, or ever will be" (Sagan, 2013). That is a very peculiar statement coming from a scientist. It has the dogmatic finality one might expect coming from a religious extremist. It certainly isn't based upon reason or logic. In fact, it most likely isn't true.

Big Bang

One of the popularly accepted theories for the origins of the universe is called the "Big Bang". Physicists have proposed that the universe came into existence in an ancient, massive, explosion of heat and light. This would appear to be a powerful confirmation of the account in Genesis where it says that "darkness was on the face of the deep . . . And God said let there be light: and there was light. And God saw the light, that it was good: and God divided the light from the darkness" (God, 2010)

So the universe not only had a beginning in space and time, but it began at the beginning of space and time. The second law of thermodynamics has long suggested a beginning. The second law, accepted and known by scientists since at least the early 1800's, simply states that all things are in the process of breaking down.

There are many examples to enumerate: old cars quit working properly, buildings decay, humans age and die. It is true that new houses are built, and old cars repaired, but these things only happen if we add materials and energy to their systems. New life is born, but again only at the expense of energy. Scientists call the increasing disorder in the universe "entropy", and the second law says that it is always increasing. That is why Bob Dylan's song, "Everything Is Broken" is so timeless.

Entropy

The sun is a classic example of entropy. At its core, it burns Hydrogen and turns it into Helium. Physicists predict that the sun will eventually run out of fuel and go cold. But the sun is just one example of many such burning balls of gas, or stars, and all appear to be running out of energy. If they all appear to be declining in the same direction, then there must have been a time in the past when all of them were initially ignited.

Then when Albert Einstein published his theory of general relativity in the early 1900's, one of the conclusions drawn from his theory is that the universe is expanding. Einstein, himself, had not foreseen that conclusion and was irritated by the suggestion. He spent some time trying to disprove that expansion because it implied a beginning.

But later Edwin Hubble observed something called the "red shift" that suggested that distant galaxies were moving away from each other, just as predicted by Einstein's theories. These distant galaxies were flying apart at incredible speeds which seemed to only be accounted for by a gigantic explosion at some point of origin. Their trajectories all seemed to be coming from a single point and time. They were flying apart so fast that the speed overcame the gravitational pull that scientists had expected would pull them together.

With this mounting evidence, scientists were forced to conclude that the universe must have experienced a creation event. It is revealing that, by sometime in the middle of the twentieth century, many scientists had become atheistic, and the hypothesis of a creation event was very disturbing to them.

For example, Arthur Eddington, an astronomer, called the theory "preposterous . . . incredible . . . repugnant." Philip Morrison, a physicist, stated publicly that he wanted to reject it. Many scientists worked very hard to try to formulate a hypothesis where the universe has existed forever, so they could avoid a creation event.

Eventually a new theory was proposed that became known as the "steady state" universe. This theory was proposed based upon Newton's theory that nothing could come from nothing, *ex nihilo, nihili.* Thus the universe could not possibly have arisen from thin air, or nothingness. In order to satisfy the requirements of the second law of thermodynamics, this theory posits that, as energy and matter burn up over time, new energy and matter is created. The universe is eternal but remains in a kind of steady state. That way the amount of energy and matter in the universe stays equal.

Of course, there is no evidence to support this theory, and it violates Occam's razor. But many scientists admittedly adopted it because it provided a way for the universe to be eternal and, therefore, not need a God and a creation. Saying without evidence, that matter and energy are eternal does not in any way explain what caused them to become organized.

However, another projection from the "Big Bang" theory was that, if the universe originated from a single explosion some fifteen billion years ago, some of the radiation from that blast would still be around. In the 1960's Arno Penzias and Robert Wilson discovered radiation coming from space. It did not seem to have an origin but came from everywhere and had a temperature of about five degrees above absolute zero. This radiation was exactly what was predicted for the residue radiation from a creation event and has since been confirmed by numerous other projects and scientists.

Scientists continue to put forth new theories suggesting that the universe is only a part of a still vaster structure that perhaps exists in another dimension. Others hypothesize that new universes are created when two super spaces collide. But it needs to be recognized that most of these hypotheses are put forward in an attempt to circumvent a creator and a creation event.

The existing data, at this time, points strongly to some creation event at a specific place and time. Those, who assume there is no God, find that unacceptable and so continue to formulate outlandish theories with no evidence to fit their assumptions.

Interpretations

The purpose of providing a brief history of the scientific view of the origin of the universe is to show how our scientific concepts have changed from time to time. Historical thought also shows how basic scientific assumptions influence the theories proposed. It is likely that much of mankind will never know exactly how the universe was created. Its very creation seems to defy the known laws of how the material world operates. In fact, those laws may not have come into existence until the universe was created. Is there more than one way to interpret the known data?

By the same token, there are multiple ways of interpreting the Bible and other scriptural sources. Consider this quote. "In the beginning God created the heaven and the earth." The Bible is not a scientific treatise. However, these words as translated from the Jewish Torah are unique in the world's religions. Most religions have dodged questions about the origin of the universe by either ignoring the subject or concentrating on just the origin of the earth, or by postulating multiple universes, creations and vague beginnings.

Only the Judeo-Christian tradition boldly proclaims that the universe had a beginning. The Bible was written well more than 2,500 years ago and, perhaps, is based on oral traditions that are even older. There was no necessity for the writers of the Torah and Bible to begin with the creation of the universe. The story could easily have begun with the Garden of Eden and been more in line other religions' origin-sources. But in Genesis, the early author boldly declares that the universe had a beginning in a particular place and time and was created at the will of God.

Genesis continues by saying that there was darkness on the face of the deep, and God said, "Let there be light: and there was light." Many critics have pointed out that God did not create day and night until day four, according to the account in Genesis; but that He created light first. This was unexplainable for thousands of years.

However, modern physics suggests that is exactly what God did; create light from a massive explosion. Today's physicists believe light was created fifteen billion years ago. But day and night would not have come into existence for a very long time after that, being brought about by the formation of the stars, planets, and the rotation of the earth.

Of course, many secular writers question the use of biblical description of creation having taken six days. However, a few years ago, I took a brief class in Hebrew. I learned very little, but one thing was clear. The Hebrew term for "day" is used as only a vague time reference. The term for creation could be equally interpreted as a season or an era. The seven-days reference is more of a sequence place holder than a specific time description. I learned that piece of trivia, of course, because we began our Hebrew study in the book of Genesis in the Old Testament.

I am not trying to prove that God exists either from science or from the Bible. What I am saying is that there is not necessarily any conflict between the scientific account of creation and the Christian account. Again, the Bible is not a scientific textbook. It does not provide a detailed account of how the universe was created. But there is also no conflict between what the Bible does say and what science has come to accept as evidence of what happened. Historically, science has seemed to be at odds with the Bible. But in an amazing turn of events, scientists have arrived at the conclusion where theologians began.

If everything that begins has a cause, then the universe had a beginning; and so, must have a cause. For several centuries, people did not believe that the universe came into existence, but that it has always been there as a perpetual motion machine. However, modern science has shown that it had a beginning, and so it had a cause. That beginning is called the creation, and the cause is called God.

Age of the Earth

There are those who insist that the earth is only 6000 years old or less. The scientific explanations offered above assume large expanses of time: literally billions of years. The earth itself may not be quite that old. Geologists are the scientists who study the formation of the earth and the methods and time frames involved.

Biologists sometimes contribute to their information by telling them when and where we think various plants and animals have lived. It was some of this kind of information that led to the theory that the continents have moved apart over time through continental drift. But it is primarily the geologists who have set the age of the earth, although much biological data seems to fit nicely into their scheme.

I've already explained that scientists don't believe that the earth was created in just six of our twenty-four-hour days. I also offered a brief explanation of the creation belief concerning the translation of Hebrew. There are other Biblical passages that are used to excuse the discrepancy, such as the verse in 2 Peter 3:8 that reads, "With the Lord a day is like a thousand years . . ."

But I have studied the Bible and other scriptures carefully and cannot find any place where it says that God was ever finished creating the earth. It is true that, in the Genesis account, He took a break after six days. But did that mean that He didn't go back to work on Monday morning?

While we may never know the laws He used to create the universe, why do we think He should be restricted in how He created the earth to only supernatural events? If He created the universe using a set of physical laws, why would He not continue His work of creation using those same natural laws? After all, when He created the earth, He said that it was "good". Why would He jump outside the use of natural laws to continue His creative work throughout the universe?

I don't declare that His creation is continuing, or that He operates only through natural law. The point is that there is certainly nothing in the Bible, or science, that indicates that this isn't the case. Once again, I hope to show different interpretations and the possibility that science and religion are not necessarily contradictory. This is maybe not as exciting and controversial as atheist screed but certainly more reasonable. Reasonable, of course, infers the use of reason.

Chapter 5 - Man's Special Place

I don't know . . .
    I don't know if man is truly the dominant species on the earth. I know that we think we are. Some think it is good that we are the dominant life form. Others think we are destroying the earth. But what if all we really are is the conscious life form that creates categories like "most dominant" and "least dominant"?
    Because, the fact is, we are not the dominant life form on earth. In numbers, insects are. Of the millions of animals that have been described, named and classified, three fourths are insects. Insects compete directly with us for space and food, and they often win. Humans have spent an exorbitant amount of energy combating insects, yet I am unaware of a single species of insect that we have caused to go extinct. There may be some, but not many.
    I don't know why humans have been as successful as we have been. We certainly don't have all of the characteristics and attributes that make other animals biologically successful. We aren't the biggest, fastest, strongest, best armored, most easily hidden, or most vicious. About all we have going for us is our brains and our social organization. Apparently that has been enough in the past, but it may not be enough in the future.

The Anthropic Principle
    Actually, it might turn out that we are kind of special after all. All those insects may just be here for our benefit. However, there are a limited number of

ways of explaining certain observations that have been made by scientists in recent years. Take, for example, the Anthropic Principle. To understand the significance of it, let me summarize scientific thought that has been developing since about the time of Copernicus.

For many years' scientists have postulated that mankind is just another animal that has evolved by chance from lower life forms. While this is a topic for the second half of this book, it has some application to our "standing" in the universe as well. For the argument has been, by some, that the universe happened by chance. According to this theory, the sun is a result of a giant explosion, the elements that coalesced to form the earth are here by luck, life evolved on earth accidentally, and humanity is simply the end product of a random sequence of events.

In the past few years, several physicists have asked why the universe operates according to the laws that it does. Why is the gravitational force exactly what it is? How old is the earth, and would it make any difference if it was a little younger or older? Beginning in 1961, Robert Dicke published a series of articles in which he maintained that many characteristics of the universe were not random (Dicke, 1965).

For example, he stated that the age of the universe had to be what it is for life to exist. If it were older, the stars would have burned out. If it were much younger, the proper mix of chemicals and metals would not have coalesced in place.

Stephen Hawking pointed out that, "if the rate of expansion one second after the big bang had been smaller by one part in a hundred thousand million million, the universe would have collapsed before it even reached present size." (Hawking, 1996) One astronomer (Smolen, 1997) has estimated the odds of all the characteristics of the universe lining up just right for our earth and human life to exist is one chance in ten to the 229th power.

In short, in numerous ways, our universe is just exactly what it needs to be for human existence. Like Goldilocks and the three bears' porridge, it is "just right".

Three Theories

A growing number of scientists have adopted this Anthropic Principle. The obvious conclusion is not acceptable to all, so many have asked how such a series of characteristics and events can be explained. Three different hypotheses have been put forth to explain these observations on the origin of the universe.

The first hypothesis is that the universe is the result of a happy accident, dependent entirely on chance. The odds for this event occurring are astronomical, especially if one presupposes an eternity of time. Some scientists argue forcibly against the idea of the Big Bang theory just because of those tremendous odds. Too, If there was a beginning, then infinite time was not available.

Others argue that since we are here, the game of chance must have simply worked out in our favor. However, this is really like suggesting that if twelve marksmen fire repeatedly at a man from close range, and all failed to hit him, that he was just lucky. After all, the man is still alive, so it must have been luck. Why would anyone suggest bribery, or a fix, or a cause?

Obviously, believing that we are here by luck, with the suggested odds against it, is just a little crazy, or completely doubtful. It is not merely that life turned up on a planet. It is that the entire universe seems to exist just so life could turn up on a planet!

There are also those who desperately do not want there to be a God. So a second explanation has been suggested: multiple universes. The theory, in its simplest form, is that our present universe represents only the current expansion. Eventually, our place will run out of energy and collapse back on itself until it generates a new universe: a yoyo-universe idea.

There is also speculation that there are many universes, each with its own set of laws. If there are sufficient numbers of universes, then it is probable that one of them generated us. Some even speculate that new universes are being born even as I write and you read. They suggest quantum worlds are routinely split into parallel universes.

There is one huge problem with these three universe theories. There is absolutely no evidence to support them in any way. It is very peculiar that logical, reasonable scientists would propose theories for which no evidence exists, whatsoever, in a desperate attempt to avoid God. In fact, why parallel universes would be preferred over concepts such as Heaven and Hell also seems most odd. It is almost as if these people live in a perpendicular universe.

Earlier, I discussed the concept of Occam's

razor (p. 37) It states that if there are multiple possible explanations, the simplest explanation with the fewest new assumptions is probably correct. All of the theories concerning multiple universes simply propose explanations that are fantastically complicated, with no evidence, when there is a simple and reasonable explanation: God.

The obvious conclusion that one might make from these observations is that, besides the universe having a distinct beginning, there seems to have been a distinct plan. The universe appears designed for life because it was designed for life. A designer universe is the evidence.

Believing this requires no fanciful theories about a multitude of universes about which we know nothing. We can account for the one universe that we can experience. Those who proposed the three theories merely wish to do away with an invisible God by proposing invisible universes.

Even scientists who are not religious often speak of nature and the material world in spiritual terms. Sometimes our existence is called sacred, mysterious, or mystical. Indeed, why do electrons obey laws at all? Why do inanimate atoms combine only in certain ways? The order and intelligence of our universe would appear to be best explained by an orderly and intelligent creator.

Man's Spiritual Place

The main world religions: Judaism, Christianity, and Islam, all maintain that man holds a unique position in the eyes of God. Each claims that humans are special, and some claim that certain types of humans are especially special.

Of course, no other living thing has the ability to counter that claim in any way that can be communicated to humans. At any rate, humanity's special position will not count for much if we disregard the physical requirements of our existence.

While it is clear enough what the material world is, it is much less clear what the spirit is. Does the spirit exist outside of the material world? We use the term as if it does. Can a material object have a spirit? We have no way of identifying that. Yet we use a material object, the brain, to think about and discuss the spirit.

The theologian is at some disadvantage in that there are less-precise definitions for the spiritual objects and forces in the religious world. For example, we often use the term "spirit" to describe the concept of the Holy Ghost, a member of the Godhead, or our own personal identity and eternal persona. It seems obvious that these two things ought to be different, but it is difficult to define one in such a way as to distinguish it from the other.

This is indicative of the difficulty of religious thought; the lack of precise definition and measurable quantities. We may speak of being filled with the spirit, or being dispirited, but we seldom speak of having half or a quarter of spirit.

Still, if the world and universe is indeed created especially so that man can exist, then man must be something special. And if the longing for different ways of expressing knowledge is desired by the great bulk of mankind, including scientists, we might assume that there is some component of existence that goes beyond materialism.

Chapter 6 - Is Man Special?

I don't know . . .

I don't know if I even know myself anymore. That may sound like a strange thing to say; but according to modern brain research, it may be amazingly true. We've known for a long time that our brains consist of specials cells we call neurons that have the ability to send tiny electrical signals. That is how our bodies communicate internally.

When you step on a bee while walking barefoot, it is important that your brain finds out about this quickly because there are numerous, coordinated responses that need to take place. Most urgently, your body thinks it needs to get the bare foot off of the painful stimulus. But in order to do this, the other uninjured foot needs to be planted firmly on the ground. If not, you could find yourself in an even worse position.

What is not clear to us is how this massive collection of billions of neurons prompts thought. Some parts of the puzzle are becoming clearer. For example, we know that the very top part of the brain, called the cerebrum, is where we are conscious of what goes on around us. The cerebrum is composed of six layers of cells and is about the thickness of a stack of six playing cards. These are interconnected to one-another in an amazingly-complex, three-dimensional, network or matrix.

At birth there are approximately twenty billion neurons in the cerebrum of a human brain. Each of these cells is highly branched with an estimated one thousand connections to other neurons each. The number of connections, then, is truly staggering.

As a baby grows, a small number of neurons, that don't receive much use, may be lost. But that number is only in the range of a few hundred thousand, an insignificant number compared to twenty billion. The cells that remain can form new branches connecting to other neurons. Some branches may be lost, and other gained, according to need.

By the time a child is ten or twelve years old they still have basically their twenty billion neurons, but now each neuron may be connected to other neurons by ten thousand connections. This, apparently, is what learning consists of; building new connections between cells.

We now know this process goes on throughout a lifetime. While it may be most dynamic in young children, humans continually form new branches and pare away old ones, even into old age. In fact, our brains change pretty dramatically throughout our lives, and we may now possess literally millions of connections that we didn't have ten years ago. We may also have lost some that we had but haven't used as much.

So you see, that's why I worry that I don't know myself at all anymore. Except, of course, I do. I am perfectly aware that I am the same set of neurons as I was ten years ago, even if I have made dramatic new connections in the intervening period of time. I am still me.

Material Affects

While, in the last chapter (see pg. 51), I explained how it seems clear that the universe has a special place for man, it is not clear to everyone that man is special.

The prevailing view of science is that the only truth is "material", because material is what everything is made of. This view comes about because science is a system devised for studying the material world. There will be more on this in the next chapter. Science more or less assumes that everything is made of materials, and so materials are everything.

It is easy to understand how this kind of thinking can exist because the material world definitely influences our lives in many ways. Let me give you a personal example. For many years, I have experienced a deep lethargy during the winter months. This is a medical condition, or disease, and is named Seasonal Affective Disorder (SAD). In fact, it is not truly a medical condition at all. It is part of the natural variation among people when they are adapting to the seasons.

There is a tiny gland at the base of the brain that holds neurons which are directly wired to the eye. This gland is partly in control of our day and night cycles. As it gets dark, the gland releases a hormone called melatonin that makes us sleepy. When you see a baby rubbing it's eyes, you know that melatonin is increasing in the blood stream. When the light increases in the morning, the melatonin levels drop, and we awaken.

As the days grow shorter, and sunlight grows dimmer and briefer from the tilt of the earth, this gland becomes more active. Some people secrete more melatonin than others, and it never entirely clears from the blood stream. Those people are left feeling sleepy, groggy, cranky and otherwise slightly sad.

Consider that if a chemical molecule can literally change my mood seasonally, perhaps my mood consists of nothing more than chemical molecules. And if my mood is a chemical molecule, then I am the result of the sum total of all my chemical reactions. This is what some scientists claim.

By contrast, I hope to show that though the material, physical world can affect us, it is not us.

Fundamentals

Fundamentals are defined as those things from which everything else is made. Since our physical bodies are made of materials such as atoms, then it follows that we are material beings. But there are other kinds of fundamentals besides atoms. In fact, there are many kinds of fundamentals.

Every field of study has its own set of fundamentals. In mathematics, fundamentals are the whole numbers. In music, the fundamental is the twelve tone scale if you study western music. In language, the fundamentals are the phonemes or the sounds of the language; and in writing, the fundamentals are the letters representing the sounds.

One of the unique features of fundamentals is that fundamentals can be combined to form new fundamentals. These, in turn, can be formed into still more structures that can also act as fundamentals. Letters become words, become sentences, become paragraphs, and become chapters, which become novels and textbooks.

In fact, conceptually, we now understand that anything can be built if you have just two opposing fundamentals. Binary code can make music, math, or the play Othello. For that matter, Macbeth could be written in Morse code with dots and dashes and still be Macbeth.

However, it would be ludicrous to maintain that the dots and dashes that make up Macbeth are somehow all there is to the play. The play obviously carries some meaning far beyond the fundamental signals that communicate it. It exists as a unique whole that is more than the sum of dots and dashes. And it has an effect on us different from the materialistic sensation of sight and sound we receive.

Man as Material

There are prominent scientists who maintain that our brains are neurons and chemicals and nothing else. If you put enough chemicals and neurons together, you get a soul. Interestingly, it is the same handful of militant atheists who make all of these arguments and blindly ignore other hypotheses.

One such scientist, Jerome Elbert, has said, "If such interactions exist [communication between soul and brain] the human brain is an interface to another, nonphysical world. Such interactions suggest the rules of science apply to all of the universe - except human beings." (Elbert, 2000)

This picture gives humans a unique position in the universe. Anthropocentric thinking regards humans as the central force or most important element in the universe. This anthropocentric picture seems very unacceptable to the scientific world view. Obviously, the brain as an interface between the material world and another non-physical world is entirely acceptable to religious persons.

So there you have it. Anything that cannot be explained in materialistic terms is not acceptable. I wonder what scientists such as Elbert do with Beethoven's fifth symphony. Which "anthropocentric view" is being talked about, the actual physical fact or the fact that our experience is more than the physical events? By stating that it is unacceptable, he is saying that it is not real. But as C. S. Lewis has observed, humans are entirely mixed up over the use of the word "real".

C. S. Lewis goes on to say:

"Human beings are completely confused about the meaning of the word "Real". They tell each other of some spiritual experience, 'All that really happened was that you heard some music in a lighted building'; here 'real' means the bare physical facts, separated from the other elements in the experience they actually had. On the other hand, they will also say, 'It's all very well discussing that high dive as you sit here in an arm chair, but wait till you get up there and see what it is really like'; here 'real' is being used in the opposite sense to mean not the physical facts (which they know already while discussing the matter in armchairs) but the emotional effect those facts will have on human consciousness."

The material world is real, but it also has effects on humans that go beyond electrons and atoms, or even physical manifestations such as force and temperature. It is true that science has no tools to investigate non-physical events, but it does not follow that non-physical events are not real. Because a musician concentrates on sound does not imply that sight is unreal. Because I use reason and logic on a daily basis does not mean I do not love my wife.

Free Will

There are problems with the scientific view of humanity operating strictly as material beings adhering to physical laws. The most pressing difficulty is the conflict with the idea of free will. In fact, some scientists declare that free will is an illusion. Francis Crick has said that while it seems free, ". . . it's the result of things you are not aware of." (Horgan, 1999) E. O Wilson writes that "the hidden preparation of mental activity gives the illusion of free will." (Wilson, 1998). We are promised that, as we learn more, we will understand these "hidden preparations" and things we are "not aware of". But that is not an explanation. That is a promise and one that they cannot guarantee to keep.

Once again, seemingly reasonable and intelligent men are inventing hidden and invisible forces to avoid conclusions they do not like. E. O Wilson and Francis Crick were both free to make whatever statements they wished to make. They are free to choose to write books, or not. There are no physical forces, hidden or otherwise, that make them say the things they say. That is the whole point!

They have chosen to say them. No physical law has prepared their minds; they could have refrained from writing anything.

If I am sitting at my desk writing, I have the ability to continue writing, or to stop. I am not compelled by any law to continue to pound out these words. Physical laws apply in every circumstance. If I drop my pencil, it will fall towards the floor. The pencil has no choice. But humans do. The computer cannot rebel against what I say. But humans can choose to run in front of a train, overriding their normal self-preservation. Some humans choose to remain childless. Others choose to have many children. There are no physical laws governing these issues.

Furthermore, humans are moral beings. We think in terms of what people ought to do and should not do. One may profess to be non-judgmental. But if I walk up to you and hit you in the nose, you will think that I "should not" have done that. Immanuel Kant observed that words like "ought" and "should" have no meaning unless they imply our ability to choose. If I have no choice, "ought" has no meaning. We are free to choose between alternate courses of action, and so we cannot be operating only under some kind of physical law. And if there is no ought, there is no law of any kind.

Consciousness

There are those who believe that our consciousness is nothing more than the hugely complex connections of the neurons of the brain. They talk about the brain "hearing" and "seeing". In this respect they are correct. The brain has five lobes, but four of them are areas where our physical senses are interpreted, not where we "think". Scientists also talk about people "feeling", "believing", "thinking", or "guessing". But a brain removed from human support, does none of these things. It is a fallacy to talk about the brain seeing. Such people are attributing human qualities to an object. My brain does not perceive or understand; I do. I use my hands to write, but I would not say that my hands are writing. I use my feet for walking, but I would not say my feet are walking.

A neuroscientist may know more about my brain and how it functions than I, but he cannot know what I am thinking. That is unique to me. I experience myself separate from the physical world and sensory information I receive. I know full well that the physical reality surrounding me is not the full story. Numerous humans have attested to this fact after suffering hardships and persecutions. The materialist forgets to account for his own knowledge and thoughts. If some scientists are correct and there is no "free-will", it does nothing to help me negotiate this life or survive in a three dimensional world anyway.

We experience other mental and spiritual characteristics and states besides consciousness. I can have intentions and feel guilt. I have beliefs and desires. I can set goals, seemingly unconnected to environmental influence, and carry through purposefully. I watch others behave in meaningful ways and interpret their actions.

When Material Changes

These characteristics are all the more amazing because my physical being is not static. The atoms that make up my body are readily replaced. Within the course of a year, there is probably not a single atom in my body that was there a year ago even though the number of cells may not have changed very much. We are much more like flickering flames than material objects.

Further, my physical body changes in other ways as I age. Yet I still recognize myself in the mirror, and people who've known me still know it is me. My thoughts and identity are still mine, although I recognize that I am thinking differently than I did thirty years ago. Yet I know I was not a different person thirty years ago.

True Belief

A last difficulty with a material humanity is that it is illogical and leaves us with no basis for action. For example, if I believe my brain and consciousness is nothing more than cells and chemicals, then I have no basis for believing my beliefs. My brain may, in fact, not be made up of chemicals. It may well be just the chemicals in my brain that makes me think it is so composed. This kind of thought leaves all of logic and science a fraud.

Chapter 7 – Objectivity and Science

I don't know . . .

    I don't know which antennae to buy. To tell the truth, I hardly understand the differences between them. See, I just got my amateur radio license (KD0LUV, thanks for asking) and a new radio. It works pretty well but, right now, I just have a little, rubber-duck antennae, and I want to be able to communicate farther away than it allows. So I am talking to some friends and reading specifications in various catalogs. Actually, it was difficult to decide which radio to buy in the first place. So I finally bought one that a friend has and says he likes. They are all electronic.

    On the other hand, it has been easy for me to decide which chemicals to use in my lab for a new experiment. I have tried several of the photodynamic dyes before and studied their various light-wave sensitivities. So for what we want to examine, and the limits of our lab, it is pretty easy to decide. For now, we're going to stick with Rose Bengal.

    But I don't know which movie to go see. There are several out that appear interesting. But I have been burned so many times by paying for stupid action movies with no plot. Or I've been offended so often by the glamorized portrayal of casual sex. And most of the time, the language used is not what is spoken in my world. I just don't know.

    I wish I knew someone who has seen the movie so I could ask them what they thought. Well, actually, I do know some people, but I'm not sure their tastes are similar to mine. I wish I knew someone who knows me well, and knows movies well, and that I trust to give me some good advice.

Scientific Truth

That is somewhat of an oxymoron. Science can contain truth, but truth need not be scientific. Science can also be untrue. Truth is true, regardless of how it was discovered. So to speak of "scientific truth" is not exactly correct.

In our technological age, somehow, science has come to be seen as a complete framework for understanding mankind and the universe. Thus, if something is not scientific, it should be automatically rejected. While this is often claimed to be the only rational view, in fact, it is irrational. It is not even what you and I normally seek in our everyday lives.

Notice that in my opening remarks to this chapter about radios, lab chemicals, and movies, I carefully chose the topics I did to make my point. In only one instance out of three, was I able to make a decision based purely upon scientific data and that is actually a high percentage. In our daily lives, we seldom desire more scientific data. Knowing the decibel level of the music does not help us choose what we listen to. Even if I know something about the complex workings of radio antennae, the decision is complicated and personal. More scientific data is seldom very useful. We almost always would rather ask guidance from someone who knows the subject, the situation, and us.

Obviously, then, there are many more sources of truth than "scientific truth". In fact, scientific truth may only be useful in narrow, special instances. In those cases, it may be very important. But on a daily basis, it has severely limited uses in our lives.

Science cannot tell me what music is best, what the meaning of a novel is, which person to marry, how to relate to my elderly parent or a misbehaving child, or even which kind of candy bar to purchase at the checkout.

Where do the other truths that seem to help us negotiate the world come from? What other kinds of truth are there? The answer to this question is probably beyond the scope of this book, but I think I have demonstrated that scientific truth is not the only kind of truth, or even the most useful kind of information we might have.

Reason and Truth

Most of the technical and specialized fields of study pride themselves on being reasonable and logical. Skilled laborers love to tell stories about the university professor who didn't know how to fix his own door. Medical doctors are continuously amazed at the ignorance lay people have concerning their own bodies. Lawyers often have trouble disguising their superior intelligence from the lay person.

Perhaps no group is more proud and self-congratulatory about being reasonable than are scientists. Science claims to go wherever the evidence suggests, even following new and unusual theories. And yet, as we have seen in previous chapters, some scientists willingly embrace theories with no evidence whatsoever and refuse to entertain obvious conclusions when the evidence does not fit their prior assumptions. While not all scientists may be guilty of this manner of thinking, some are, and it seems peculiar.

Science seems to have changed dramatically in the last one hundred years. Since the time when most scientists thought they were "discovering the mind of God", many of today's scientists actively prohibit the idea of God. Today's science is based on the fundamental premise that nothing is supernatural. One scientist, paleo biologist Douglas Erwin, has stated, "One of the rules of science is, no miracles allowed." This is only one statement of thousands I could cite to show that modern science has been designed to exclude a designer.

But is this stance reasonable? Why is it necessary to object to theories because they imply a beginning? Why is it important to avoid the special position and attributes of humanity? If the evidence suggests a beginning and a designer, why not test that theory? Today's scientists routinely accept the orderliness of nature and use basic laws that seem universal, yet they never question where the order or laws originated. Scientists tend to reject any suggestion that there might have been an ultimate cause of otherwise unexplained events.

For example, it is generally accepted in the scientific world that living cells arose from inanimate matter by some kind of evolutionary process. This is not a statement of fact. It is an assumption. All reason is based upon assumption. It is hard to understand such a proposal unless the purpose is to specifically exclude a creator.

Past scientists have seriously suggested that the first cells were brought here from another planet by way of some meteor or other heavenly body. There is no evidence for this, and it dodges the entire question of how living cells came to be since it only moves the question to an unknown site in the universe.

The initial assumption of modern brain scientists is that consciousness is a result of a peculiar way neurons function, that consciousness is basically an illusion. What, to me, is a startling conclusion might be another's beginning assumption.

Yet intelligent, educated, reasonable men make these theories and actively promote them to the public through their books and lectures. No explanation of anything that makes reference to God, a beginning, a plan, or a designer is to be allowed. This level of dogmatism is astounding in otherwise reasonable people.

How can such closed mindedness come to be? I believe there have been two forces that have contributed to this significant shift in assumptions. I am not sure which is the more important, so the following proposed explanation is not necessarily in order of significance.

Secular Education

Beginning in the eighteen-hundreds, the model of the German University with its nineteenth century advances in scientific understanding, began to be influential throughout the world. There was no universal education system in Europe or anywhere else in the civilized world at the time. So the advantages of such an education were severely restricted to the more well-to-do families.

But as more and more people became educated, there was often not much for them to do in the culture. The universities began to admit others in an attempt to expand their influence and to enrich society as they supposed. As more people became educated, the universities began to operate separately from religious influence. Education, and especially higher education, became increasingly secularized. Secularization is the transformation of a society from close identification with religious values and institutions toward nonreligious, or irreligious, values and irreligious institutions.

Then in the early twentieth century, the idea for a universal public education was initiated. The basis for beginning such an effort was threefold. Partly it was seen as an economic need driven by the industrial revolution that needed workers trained to work in factory settings. It was also justified by the claim that democracy required an educated and enlightened citizenry. It may also have been driven, in part, by the seemingly universal need parents feel to better the lives of their children, or to be the architects of their children's lives.

There really was no evidence that public education would improve people's lives directly. The level of literacy was said to have been extremely high in the early colonial days of the country without such a universal public policy. In fact, the idea resulted in almost immediate conflict between school needs and the need for a child to be at home on the farm.

As academic achievement and literacy were stressed, and as teachers became more educated in the secularized universities, instruction in religious subjects, principles, ethics and moral behavior declined. Over time, even the study of the great philosophical ideas upon which the United States was founded, became less common in the school curriculum. Further, the time spent learning school curriculum took away from time spent learning the basic and practical skills at home. The implementation of public education began a trend that continues today with more and more of children's time being spent learning school curriculum. As a result, children are almost completely removed from every day, practical experiences.

Where and when do children today learn to plant, sew, splice a wire, keep bees, care for animals, and be responsible for other people? Where and when do they learn to cook, build, repair, change a tire, repair a broken door, and hosts of other skills that have sustained humanity for thousands of years? Practical application of these fundamental skills has been wiped out in a short 100 years. It can be effectively argued that many of these practical skills led to the inventiveness and creativity of Americans in the past.

Consequently, this whole trend has led to increased secularization of the educated class. Those who persevere through many years of higher education are especially affected. The long indoctrination into subject matter, to the exclusion of other types of education, has created entire fields of specialists. These people know a great deal about one subject but very little about other fields. More importantly, they don't know how to do much of anything useful.

Secularization of education continues unabated and is reaching ever lower into elementary grades. It is not so much a problem of science or theology. It is a problem of the people. People have come to view education as having everything to do with economic benefit instead of practical abilities or moral persuasions. It is a benefit for many families to have others educate their children while parents engage in distant work. Of course, the goal is to earn greater wealth from their specialized education that will, in turn, benefit the children.

There really is no evidence that nearly-universal literacy has made people wiser, happier, more free, or for that matter, even wealthier. Improved economic standards correlating with increased public education does not necessarily show causation. Public education certainly has changed how we live, but the assumption on which it was begun was based upon economic wellbeing. That it has, or has not, worked remains an assumption. Without question, though, it has resulted in an increasingly-secularized, narrowly-educated citizenry.

Evolution

Evolution is a topic for part III of this book, but I must mention it here because of its dramatic effect on science in general. The Darwinian Revolution cannot be called simply "another scientific theory". Evolutionary thought created a new world view. Prior to Darwin, the idea of the supernatural was accepted as a real and possible explanation of unknown events. The theory of evolution made it possible for previously-unexplained mysteries of God being given scientific explanations, even if the ideas were flawed. Henceforth, science became increasingly secular, and scientists became increasingly unwilling to consider supernatural explanations. In our present world, this philosophy has reached an extreme where reasonable men have become completely closed-minded to any value of spirituality or nonmaterial causes and effects.

Oddly, secularism sometimes harms science. I believe many scientists recognize that, although often not consciously. How is secularism harmful to science? Science has cut itself off from other means of obtaining truth and experiencing the world. This increasingly narrow way of interpreting the world harms science in two ways.

How Secularization Harms Science

Many students today avoid science. I have been teaching science for more than forty years. There are increasing numbers of students each year who find the perceived battle between science and religion a hurdle to be jumped in their pursuit of a career.

They offer many different reasons for feeling this way. Some students come from close families with strong religious beliefs. They receive pressure concerning scientific studies and simply decide, consciously or unconsciously, they would rather remain close to their roots. This does not make them dumb, short-sighted, or even anti-intellectual. They just perceive that there are many ways of making their way in the world, and they choose one that brings them more satisfaction concerning things that matter to them.

Of course, there are also students who are simply put off by the narrowness of science. When everything must be seen through the lens of materialism and nature, and when their every experience tells them that there is much more to the world than that, they simply decide that while science is interesting and useful, it is just too restrictive.

These students are capable and intelligent people. They simply feel constrained by the field. Often these students study science to open the door to other fields they see as being more practical and less restrictive.

In both cases, science is deprived of many intelligent, capable, even brilliant students because of scientists' inability, or unwillingness, to look at the world in any other way. Of course, there are many scientists who are not so closed minded, and many who are even Renaissance in their abilities and world view. But they are not the public face of science. It continues to be interesting how many scientists recognize the intellectual straight jacket they are in and struggle to break out of it. I know numerous colleagues who develop outside activities and hobbies that are often far removed from science.

There are those who play in the symphony, others who paint, and some write. I have a friend who has become a regional expert on the history of local Indian tribes. Some work in non-profit, community organizations. In conversation, they all recognize that there is more to life than data. They find reason and logic tiresome and limiting after a while so develop other pursuits to make up for the shortcomings of their own scientific world view.

The Advantages of Secularism

In other ways, science benefits from secularization. It is the job of science to investigate the natural, material and physical world. In the domain of "natural", the supernatural is excluded. That is reasonable. It is not because the supernatural doesn't exist but because science is in the job of searching for natural causes.

That means, whatever subject a scientist wishes to study, must be studied from a material point of view. If scientists are to study or discuss the soul, or consciousness, or origins of man, they must look for material evidence for study because that is all they can use. If they can find no material, they cannot study it. Unless, of course, they can find material effects of immaterial things.

And that is exactly what I want as a scientist. I do not want my colleagues to end their scientific papers by saying they couldn't find what they were looking for. They are totally stumped by unexpected experiences, and so they are calling them miracles.

The search for natural explanations has yielded some impressive results! We cannot even be sure that, at some later date, someone else might not find a natural explanation for the presently unexplained phenomenon. For right now, we don't have to believe anything based only upon some possible future research. But it needs to be recognized that this is what scientists do. They look for natural explanations based on physical material to explain our natural and material world. But to state that science does more than this, would be dishonest.

For example, how would one answer the question, "Why does a frog jump?" If you are a physicist, you might suggest something about the mechanics of the frog's legs as levers or the necessary force it takes to overcome inertia. If you are a chemist, your answer might be something along these lines. The proteins in the muscle cells react in such a way as to ratchet together and make each cell shorter. If you are a child, you might explain that it jumps to escape your clutches. If you are a storyteller, you might imagine that it is to escape the boiling water. There are many truths.

Stories

"There is an ancient mariner, and he stoppeth one of three."

"Once upon a time, in a galaxy far away. . ."

"Call me Ishmael."

"It was a pleasure to burn."

These are all first lines from famous stories, great stories! I have read them all. At times these fictional stories seem real. I guess they could be real, for all I know. In fact, having dabbled in writing stories myself, and having read what a number of authors say about their stories, I suspect some of them ARE real. That is, authors take events that were real and change them, embellish them, dress them up and make them unreal; but, at the same time, they sound real.

As I said, though, some stories are real. True stories are often more compelling than fiction because we believe in our hearts that they really happened. We are filled with empathy for the characters involved in our reading. Yet, if you think about it, even true stories are not entirely true. Parts of true stories are probably even untrue. When my Father used to tell fishing stories, you could usually be sure that he really did go fishing about the time he said he did, and at the place he said he went. Everything in between might be accurate to some degree, or not at all.

People often use stories to prove a point. Stories make great analogies, metaphors, examples and parables. Politicians like to tell stories about poor, suffering citizens who will benefit by the bill they are sponsoring. People respond to stories more readily than they do to mere logic. Story telling is probably our natural mode of thinking.

But just what is a story? A story is the recounting of a series of events. The telling of an event is speculative and cannot be proven. In fact, most of the time we wouldn't even think of trying to verify every element of those events. We simply know they aren't literally true. Even if we suspected they might have an element of truth, we would recognize that it would probably be impossible to prove what was true and what wasn't. After all, the story simply arose out of someone's mind.

Many theories scientists have constructed about the universe and humanity are really nothing more than stories. While the facts that give rise to a plausible theory may be indisputable, the theory itself is a "story": a series of speculative, imaginary events used to explain how something might have happened.

Let's examine the story of "Oliver Twist". Dickenson was imagining how things might have been in early19th century London. In fact, he had seen some of it personally. What he wrote about the time might be authentic, or not. It might contain true elements that are embellished or downplayed. Perhaps the work houses were not really as wretched as he portrayed. But, in no way, does his writing represent science which is based upon evidence, proof or the ability to prove something false. We can never know if work houses were as bad as, or worse than, Dickenson claimed. Perhaps they were really a mixed bag of conditions. Because the events cannot be replicated, observed, or verified in any way, theories can be accepted with no greater weight than any other story.

Thus we can say that theories are merely stories. These also take upon them the same characteristics as any set of Articles of Faith, and the story is simply adopted on faith. Those who choose to believe may belittle and ridicule those who do not, but they have no special standing to do so. Those who choose not to accept the story are not on unequal footing, although they may feel the need to come up with a viable alternative story to account for the facts that led to the theory. Unfortunately, mocking and ridicule have become tools used by some scientists and theologians.

Who Is Reasonable?

Theologians are sometimes willing to condemn whole categories of people - literally to Hell - based upon minute details of doctrine. They are sometimes called dogmatic and closed-minded. I suppose a popular perception concerning those who are religious is that they are hypocritical, bigoted, fanatical, and ignorant.

In fact, theistic people are more open-minded in some ways than the atheist because they do not deny the existence of the natural and material world, nor the findings of scientists who provide good evidence. They do not deny electricity, gravity, or the heliocentric theory of the universe. They accept that there is a natural world that can affect them in many ways. The theist must keep an open mind about many subjects and be willing to change them as the evidence suggests.

But just because science cannot come to any conclusion about the soul, a deity, or the purpose of life does not mean that we might not correctly come to conclusions about these and similar matters using other forms of truth and experience. There are more kinds of truth than scientific truth. Science cannot say anything about any subject that is not natural and/or material. Science, itself, is not a source of atheism. It is simply used by some atheists to promote their own ideology.

Chapter 8 - The Limits of Science

I don't know . . .

I don't know why my wife and I had to lose babies through miscarriages. I don't know why some people suffer mental illnesses. I don't know why there are natural disasters. I don't know why some people in the Islamic world hate America and would kill thousands of innocent people.

I have written poems about these things. I have written short stories and unpublished novels. I have railed against greed and power. I am appalled at the living conditions in our cities. I maintain that cities are the result of driving people off the land through corporate farming that works more like strip mines than farms.

When I was young, I didn't understand the Viet Nam War, so I went to some demonstrations against it. I sang protest songs in coffee houses. I argued with friends about the need for the war and the wicked, corporate, military complex that supplied it. But the war didn't end, and I ended up in the Army.

All my words, music and demonstrations didn't save a single life, feed a hungry person, cause even one Muslim to like the US, or comfort someone who was suffering from dementia. It all goes on and on, and I don't know why.

Suffering and Science

Perhaps you think that suffering should not be a topic for a book about religion and science. Suffering has long been recognized as a legitimate subject for a discussion in religion. Job, of the Old Testament, asks the eternal question about why good people must suffer. Well, I suggest that this is also a good question for scientists on several grounds.

First of all, science inserted itself into the question when it began to explain what had been thought of as "supernatural", according to natural law and material characteristics. Historically, people did not know about causative agents of disease such as bacteria and viruses. The common explanations for diseases usually involved either witchcraft or sin. Often those who fell ill were suspected of having secret sins. But now we know that there are biological causes for illness, and disease is not caused by the hand of God or witches.

Secondly, most of what we call suffering can be seen as a physical, material event based upon natural laws: disease, disasters, and accidents. In fact, science likes to portray itself as a discipline that seeks to lessen suffering. If science wishes to take credit for curing a disease, predicting a violent storm, or making for a safer world, then it seems it should have some responsibility to speak to suffering.

Scientists have no trouble asking for public funds on the basis that they will be used to improve the world. Likewise, they should accept some responsibility when hoped-for improvements fail and people suffer. In fact, in many instances, science aggravates problems by intervention. For example, the attempt to eradicate malaria in the 1950's led to massive starvation from over population. Introducing rice to areas of the world with poor nutrition increased and spread Malaria.

Yet another reason science should speak to suffering is that science causes a huge amount of it. Almost every solution to a problem offered by science causes some other problem. The invention of plastics has been beneficial to the world and polluted the world. The discovery of pesticides and antibiotics has saved lives and caused pollution and resistance. The discovery of nuclear energy has created atomic bombs. Science is not always the hero.

There is yet one more reason why science should be discussed in connection with suffering. Scientists are some of the very people who raise questions about God's role in suffering. You have heard the argument that goes something like this. 'God, by definition, is all good and all powerful. Yet evil and suffering exist in the world. So if God could stop these, but doesn't, then he is not all good. If he can't stop them, then he is not all powerful. The world is full of evil and suffering, therefore God must not exist.'

This is one of atheisms more compelling arguments for many people. I have shown, in chapter 3, that science has become atheistic in practice. Science takes on a central role in the argument about the existence of God. In fact, atheists tend to use science as their prime evidence and base of argument against the concept of God. And since science has de facto accepted the atheists' position, science has taken center stage in a debate in which it is ill equipped to participate.

Definition of Suffering

To understand suffering's role in both religion and science, it is important to understand that there are several kinds of suffering. Often we don't distinguish between them. There is suffering when someone experiences true physical pain. When my son had a broken arm, he was in pain caused by a true, material defect.

When a natural disaster kills thousands of people and destroys infrastructure, those who are left behind are uncomfortable and in danger. What they experience we call "suffering", although many are not in physical pain at all. And then there is the suffering of the heart that is experienced by those whose loved ones die or are harmed from disease or disaster. Watching another person in pain brings about suffering also, but these examples are not all the same things.

The role of religion and science in each instance is probably different. Science was instrumental in relieving my son's pain from his broken arm and promoting healing. I may observe here, also, that the physician first added a great deal to my sons' pain by manipulating while diagnosing the appendage.

Science was even helpful in diagnosing why the arm broke. He had a bone cyst. However, science could not explain why there was a bone cyst in the first place other than to say that it happens in a percentage of cases out of thousands of people. On the other hand, science did very little for me as I watched him suffer. Prayer, however, helped me stay calm and was likewise helpful to my son. Science may help explain suffering, but it offers no comfort for the heart.

Science often fails in preventing disasters. Sometimes it can provide humans with an advance warning. It may play an important role after the disaster in rebuilding and treatment of the injured. LIKEWISE, amateur radio often plays a role in communications during such events. But those who survive must still endure hardship and danger for periods of time.

During these events, it is the act of people caring for each other that makes an immediate difference. Religion plays a serious role in educating people as to their proper roles and attitudes of compassion in preparation for these events. Religion may play a role in teaching people how to be prepared, and organized religion may even have prepared emergency plans for such situations. In this type of suffering, religion and science probably play somewhat equal roles, and they probably utilize each other in doing so. People use science to serve religious ends, and religion as a vehicle for the delivery of science aid.

During heartfelt suffering by the survivors of disease and misfortune, science has nothing to say. It does not necessarily comfort me to know the cause of my Father's death. It does not comfort me, console me, encourage me, or direct me as to what I should do next. How should I behave, or what should I believe, when I am told that the earthquake was caused by a fault line?

Science is simply silent in things of the heart. I have noticed that, after events that cause suffering, when there are public gatherings and memorial services, there is no talk of science. This kind of comfort and help comes almost exclusively from other people and God: meaning, comfort, divine mercy, and spiritual strength.

Cause and Effect

Cause is among the most difficult concepts to understand in science. For example, it was clearly my wife's fault that I sliced open my hand while working on the air conditioner. If she hadn't scheduled a party for 21 little girls at our house, I wouldn't have been on the roof trying to get the air conditioner operating in such a hurry. The wet roof, the slick shoes, working too fast, and general physical ineptness had nothing to do with it. My wife caused the accident.

Determining cause haunts scientists. Cause is a surprisingly difficult concept to prove. Even when the accident was clearly because driver A ran a red light, there is always a reason why driver A ran the red light, and a reason why driver B was in the intersection. What was the real "cause"? Children quickly learn to use this slippery concept when they claim, "He hit me first", or, "I didn't turn my homework in because the dog ate it." It's never anyone's fault.

The determination of cause is a difficult problem when considering suffering because of the different kinds of suffering. There are various exposures to risk, and the mechanism that connects exposure to outcome is often not known. Consider the effort to prove that cigarettes cause some forms of cancer. It took a long, long time to prove even though everyone was pretty sure that cigarettes were a major culprit.

Who Suffers and Who Causes It?

Science is a relatively recent phenomenon. The study began perhaps five hundred years ago, but it did not really make significant contributions until the last one hundred and fifty years. Recorded history dates back at least six thousand years. That means science has only been influential in our thinking for less than one tenth of the human experience. So what did humans think concerning the meteors, earthquakes and tsunamis, droughts, disease, and all manner of afflictions prior to our scientific explanations?

As described previously, humans have long attributed unexplained events to the mischief of capricious spirits, to the intervention of numerous gods, to witchcraft and other forms of evil, or to punishment for sins. These sins could be either collective or individual depending on the nature of the catastrophe.

Consider that the germ theory of disease didn't come into existence until late in the 1800's. Before the understanding of the germ theory of disease, suffering was associated with sin, evil spirits, or the gods. These concepts had been perpetuated for centuries and help explain the Book of Job in the Old Testament.

Today, many still see suffering as coming from the hands of God or some evil force. Some believe it is dealt out for misbehavior and withheld from the righteous in spite of centuries of evidence that this is not true. In fact, the accusation by atheists that God could, and should, prevent suffering is an assumption on the part of irreligious people to impose on God their understanding of what a God could and should do.

The central question raised by Job's story has traditionally been, "Why do bad things happen to good people?" However, making the assumption that bad things do happen to good people creates an informal fallacy, or straw man, since bad things happen to everyone, good and bad alike. Everyone gets sick, everyone eventually dies, and natural disasters destroy many people of all nationalities and cause suffering for all. "Bad people" suffer just as much as "good people". It's that we usually consider their suffering "just desserts". In these cases, no why questions are raised.

The more legitimate question would be, "Why does God allow suffering at all?" The answer to that question is found in science itself! The natural and material world is a closed, finite system. That is to say, there is limited space and limited energy for supporting life. I suppose there is even a limit on time since science tells us that the sun will burn off all the earth's water in about a billion years or so.

In a finite system, there are only two possibilities for living things: a small number of living things that live forever, or a larger number of living things that live shorter life spans so the energy and space can be conserved. It seems obvious that our world exemplifies the latter.

When we examine the types and causes of what we normally call "suffering", we see that suffering is defined basically as one of the following: the loss of life, the pain sometimes involved or assumed in death, the danger of losing life, or the loss to death by those people who cared for the dead person. Suffering, then, is another way of discussing death: fear of death, pain of death, loss from death, or even risk of death.

In our modern world, we have expanded the concept of suffering to include all types of discomfort or "bad" things. However, when we observe anything that isn't a matter of life and death, suffering is generally not taken seriously and is even discounted in less serious situations. With the advent of psychology, many of these non-life threatening forms of "suffering" have taken on more significant meaning, whether justified or not.

Broken hearts may feel like suffering, yet most of us realize they are not. However, abandonment of a family causes not only heartbreak but loss of physical wellbeing and danger too. Are divorces considered a cause of suffering? Disappointment in not getting a job can be confused with suffering. But everyone would say that starving to death from unemployment causes true suffering.

Death is an inevitable part of our finite, material world, and our earth could not exist in any other way. To blame God for suffering, would be to blame God for creating a finite world. But an infinite world would only be good for the few living things that got to live here. The laws that scientists accept on faith as being universal and eternal are the very cause of suffering.

The next question is "Does science eliminate death?" Of course it does not. Doctors do not, technically, save lives. They postpone deaths. Even as I pray for someone suffering from disease to be cured, I know that they will not be spared death in the long run. The suffering, of their death and my loss, will only occur at a later date. Science cannot, through its laws, create immortality in a finite system. So should we blame science for not stopping all suffering through death?

If there is a God who created the universe, the earth, and the eternal and universal physical laws that govern both, can He be considered to be the cause of suffering because He created a finite condition? Does that thought leave us blaming God for suffering and evil, or can it be some kind of evidence that He does not exist? I don't think so. It simply makes the statement that we do not yet understand why He created the world the way He did. Perhaps He had an infinite number of living things He wanted to create and needed a revolving system.

But let's go with the conclusion that God causes our suffering. What hypothesis might account for that? What motivation could He possibly have? Perhaps He didn't want to be concerned in day-to-day operations because He is busy making other worlds. Perhaps he wanted to create many living things and give them all opportunities to experience life, so a finite world served his purpose. Once again, the point is that there are many different plausible explanations.

We are left with the same reasonable conclusion that Job, in the Bible's Old Testament, came to. We do not know God. We have not created a universe. We have not created life. We are not eternal. His laws are not our laws. His ways are not our ways. And His time is not our time.

Conversely, what we call "suffering" may not be suffering to God. He may have purposes which we do not understand, and suffering to Him may not have the significance or meaning we attribute to it. To assume that because we see it as suffering it is, assumes that we know everything there is to know. Admittedly there are those who think exactly that.

As mentioned earlier, the creator exists outside the creation. I do not cause the events in my novel in the same way I cause danger to others when I drink and drive. Human hubris has been a problem throughout history. Men assume they know what God knows and see things as God sees them. We simply do not.

One last thought on the cause of suffering. We only speak of suffering being caused by God in certain specified circumstances. Usually these are concerning justice, when bad things are befalling seemingly good people. Yet in other contexts, we would never wrestle with such a moral dilemma. When we are at war, and our comrades fall at our sides, we blame the enemy. God did not fire the sniper bullets that killed people on a college campus. An evil person did. God did not blow up an office building or fly the plane into a building to take life.

We can argue about terminology. Were they evil, insane, hatred-filled, satanic, or crazed? We can wonder about why they were the way they were. Were they bullied, were they mistreated, were they mentally ill, or were they brainwashed? But in any of these cases, it doesn't mean that God made them do it. In fact, maybe it was Satan. Most usually it was other men. Let's give credit where credit is due.

Another interesting observation is that often we do not use the same kind of thinking when discussing happiness or good fortune. When a person is successful, or seemingly blessed, we say that they are self-made. Admittedly, there are those who will acknowledge they have been blessed by God. However, we seldom blame God for making us wealthy. On the other hand, we may question why certain people who are distasteful to us, seem to do so well.

Who is Compassionate?

It's true that suffering is a serious topic for religious discussion. Why do so few people seem to notice that it is a serious problem for science as well? Suffering is a part of the natural and material world, and science plays some role in the discussion. But while science can play a role in alleviating some kinds of suffering, it has absolutely no way of helping suffering of the heart, soul or spirit.

For example, one atheistic scientist proclaimed that, "The universe we observe has precisely the properties we should expect if there is, at bottom, no design, no purpose, no evil, and no good, nothing but blind pitiless indifference" (Dawkins, 1995). This speaks of utter hopelessness, meaninglessness, pointlessness, and despair. There is no comfort given here for a grieving mother, a frightened child, or a despairing friend.

Some people ask if it is not better to accept harsh truth than to believe in fairy tales. I suppose it would be if, in fact, the argument about suffering had proven that God does not exist. But as I have tried to show, it does not provide proof. It simply proves man's limitations of understanding. There are numerous hypotheses concerning why suffering might exist. Until those are disproven, or a better demonstration of God's absence is found, atheism is not proven. Therefore, I am not compelled to accept atheism only to accept the limits of my reason.

In fact, religion is the source of compassion and comfort for those who are suffering. Religion can not only use science to alleviate suffering, but it can offer hope, consolation, and understanding to those who suffer. These are things science cannot do. God is not indifferent to evil. God has enacted laws against man's evil tendencies. God has sent an emissary to the world who also suffered unjustly, who understands, who has been part of the material existence, and in the end who overcame physical limitations.

Are suffering and death the final chapter? Some scientists say that they are, but they cannot prove that. Christianity claims that they are not. Christians claim that the way to triumph over evil is to live again in a place where evil does not exist. Christians claim that there is such a place. It is part of God's plan to allow us to return there after we have learned what He needs us to learn in this material world. If this is true, then God has shown that suffering and evil are transient, that he cares greatly for his creations, that there is a plan and design for living things, and that there is a way for us to overcome the seeming injustices of this world.

The Cause of Suffering.
And of course, there is the fact that science has caused some of the most extreme suffering of all. Science fuels the weapon industry. Science contributes to the destruction of the environment and the pollution of the world. Science serves the drug industry. Science has caused fire, poisoning, electrocution, disease, and innumerable other human afflictions.

Yet no one seems to decry science as a cause of suffering. No one thinks we should stop doing science or put a halt to exploration of the universe to avoid further harm. No one ridicules science as a tremendous philosophical mistake because of the atomic bomb, or industrial pollution, or even climate change. No, the cause of suffering isn't science. It's the people involved.

We invariably blame the people who misapply the science, not the scientist. But when it comes to the natural world, we skip the people involved who may well have contributed. Instead we leap to blame God. Of course, men may have built homes and factories on the edge of the ocean or rivers, been negligent in safety precautions, followed misbegotten ideas, used the environment abusively, or many other contributing factors. Blaming God is a hypocritical, contradictory, and an impossible standard of the application of suffering.

PART 3 - INTRODUCTION TO BIOLOGY AND EVOLUTION

Chapter 9 - EVOLUTION

Motives

Someone once said that the first casualty of war is "truth". Wars exist for reasons, but the stated reasons may not be the real reasons. The supposed war between religion and science is no different.

Nowhere is war more evident in our day than where ethical questions are raised by advances in the field of Biology. In Part 2 of this book, I intend to discuss the creation of life from both a religious and a scientific point of view. There is the relatedness of living organisms, how living things change, the nature of nature, and the interpretation of the world by theologians and scientists.

What these issues all have in common, biologists claim, is the concept of evolution. However, there are serious questions about the usefulness of evolution as the central tenet of the biological sciences.

There are also more ramifications to be discussed than just man's beliefs. The ethical problems posed by cloning, genetic engineering, transplant technology, aging, abortion, and medical treatment all force a discussion between religion and science in a new and more productive way. While I cannot answer the many questions, I hope to pave the way for better dialogue. Perhaps I can advance truth and understanding.

First we must recognize that not everyone in our world wants peace and harmony. We must also realize that there are different motives for different attitudes. A group of people may all oppose the placement of a lamp post. While they agree that the lamp post must be taken down, their reasons for wanting to do so may be numerous.

Some may want a darker street to enable darker deeds. Others may want to avail themselves of the metal, wires and light bulbs of the post. Others may be irritated because the light shines in their bedroom windows and keeps them awake. None of them are asking if the light itself is of value. It is wise to be attentive to the reasons for people's motives, both theological and scientific.

Ignorance

Let's begin with ignorance. As knowledge has increased in the world, it became compartmentalized. It is very difficult for anyone to be a Renaissance man today. Seeking specialization of knowledge begins early in a person's life, although not in the same way or with the same emphasis in every culture.

As a general rule, Americans do not excel in mathematics. That doesn't mean we don't have brilliant mathematicians, or we don't have a large number of people who can do simple checkbook kinds of math, or Americans can't do special calculations related to their employment. But math has been one of our weaker subjects for years, starting in our elementary schools. This has been the case for a variety of reasons, some of which we probably don't even understand. Then, the number of

97

people taking math classes steadily declines after elementary school through students' remaining years of public education.

Without a math background, it becomes very difficult to excel in the sciences.   By the end of high school, two camps have been established: the scientists and the humanists. In college, except for a token course or two, students can pretty much exclude science, or the humanities, depending on their preferences.

For those who seek advanced degrees, the ignorance of the other field is even more highly perpetuated. One can be awarded a Doctor of Philosophy degree with never having taken a philosophy course. One can obtain a terminal degree in the arts having taken no more than high school biology.

Of course, between the practitioners of religion and science, the gap may be even larger. Many scientists I know were never exposed to religious instruction, even as children. If they were, that exposure dwindled and disappeared sometime after about the age of ten.

A friend of mine, an artist, has pointed out to me that most people in school stop drawing around the age of ten. Therefore, as adults, they still draw like ten year olds. The point is that a ten year old's understanding of art, or theology, combined with myths and impressions gathered from our pop culture, does not foster deep thinking on the subject.

On the other hand, I have friends who sent their son to a church-sponsored school at the age of fourteen. They fully expected him to study Latin,

Greek, and Hebrew to become a minister. No surprise. That is exactly what he did.

Because science requires a peculiar way of thinking and analyzing, even those who take a college-level course are not completely exposed to how science is done. It is especially difficult, with a seminary background, to fairly assess scientific claims. I suspect a great deal of the so-called war between religion and science exists because of limited knowledge of both groups.

Promotion

There aren't many stories in the news about two people who agree with each other. There may be literally millions of happy marriages, but one would never know it from the news. In fact, if two people disagree, but manage to settle their differences amicably, that doesn't make early-morning, five o'clock, or even local news. Conflict sells.

A recent example is the debate in 2013 over gun control and the second amendment right to bear arms. While there have been several high-profile murders in schools in recent years that get tremendous coverage, the actual instances of school violence have decreased, and the number of gun-related crimes have been decreasing for several years. Ironically, too, the worst gun-related crimes appear to occur where there are the strictest gun-control measures in place.

There are numerous instances each day in the United States where armed citizens help stop crimes. However, these go mostly unreported in the news, and the average citizen thinks gun violence has increased because of the tendency of the news media to milk violence and conflict for viewers.

Many individuals take advantage of the attention conflict gets to sell their products. The singer with the most outrageous behavior, costume, or song lyrics profits over others even when musicianship is lacking. Artists often try to portray the most flamboyant topics. Authors often become famous for outrageous plots, gruesome murders, foul language and other extreme portrayals.

Amazingly, some rather-bright scientists seem to have discovered this technique as well. In the last twenty or thirty years, a type of militant scientist has made outrageous, unscientific, and illogical claims against religion in order to sell books.

They have, for the most part, been rather successful. Of course, poor behavior and promotional activities are a human attribute not confined to one's career or major interests in life. Likewise, there are ministers who are willing to denounce science in the strongest of terms to appear courageous and to stir up interest in their own preaching, church related activities, and publications.

Now, if you are in the business of selling the news, these are the people you can use to help you sell your product. Many news magazines, newspapers, and television shows, concerning the so-called conflict between religion and science, routinely run in the days preceding Easter and Christmas celebrations.

Think about this. If I have devoted years of my life to a particular subject, and if I make a living using that subject, I may not actually be open-minded concerning alternative explanations. Combine that with one's lack of exposure and experience with "the other" subject, and much of the conflict becomes understandable.

Unholy War

This tendency on the part of the media to report and dramatize conflict and shock value is used by many people to further their own agendas, gain attention, and to promote themselves or their products. Not just the news, but anything of shock value, can be utilized to make money or sell ideas. Scandal, sex, and violence in movies, novels, radio, television, live entertainment and advertising are used throughout the media. However, when science uses such methods, it ceases to be science.

Unfortunately, there are those who believe wholeheartedly in the war between religion and science and seek the destruction of one side or the other. They often stand to profit in money, power, and influence from such promotion. I devote the last section of this book to these blatant, intellectual "terrorists". However, I need to make a comment concerning their tactics at this stage of the discussion.

Scientists, along with others who use science in their arguments, seem to adhere to this group more than religious persons. I am not aware of any individuals, or religious groups, who are not willing to use the advances of science to better the lot of mankind. Religions almost always embrace those things that truly absolve suffering. Obviously, there are exceptions found in some fundamentalist groups.

On the other hand, there are scientists who wish to see the complete destruction of religion.

This is not my opinion. If you want to read what they say in their own words, here are a few of their books. "The God Delusion" by Richard Dawkins, "the End of Faith" by Richard Harris, "God: The Failed Hypothesis" by Victor Stenger, "God Is Not Great" by Christopher Hitchens. Their titles exemplify their positions. Militant atheists and scientists have definitely declared war on religion.

"How can we ever know how many children had their psychological and physical lives irreparably maimed by the compulsory inculcation of faith?" - Christopher Hitchens

"Faith is one of the world's greatest evils, comparable to the smallpox virus but harder to eradicate." - Richard Dawkins

". . . faith seems to me to qualify as a kind of mental illness." - Richard Dawkins

"I personally feel that the teaching of modern science is corrosive to religious belief, and I'm all for that. . . . I think it may be the most important contribution we can make." - Steven Weinberg

"(Science) is not to provide the public with knowledge of how far it is to the nearest star and what genes are made of. . . The problem is to get them to reject . . . supernatural explanations of the world, and to accept a social and intellectual apparatus, science, as the only begetter of truth." - Richard Lewontin

"If religion . . . can be systematically analyzed and explained as a product of the brains evolution, its power as an external source of morality will be gone forever." - E. O Wilson

These are not statements of fact, logic, reason, or science. These are statements of opinion that are loaded with value-laden expressions, emotional assumptions, and deep animosity! Examine the language carefully. Words like "maimed", "evil", "mental illness", "corrosive", and "begetter of truth" are not measured and reasonable language.

Perhaps most telling of all is the last statement of purpose by Lewontin. "The problem is to get them to reject . . . supernatural explanations of the world, and to accept a social and intellectual apparatus, science, as the only begetter of truth." I mention this here simply to illustrate their use of conflict and shock value in gaining sales of their books, notoriety, and influence. It is hard to find anything akin to it in theological literature.

Back to Evolution

It was necessary to demonstrate that these religion and science discussions usually do not start with a common goal. They leave little hope of reaching understanding or furthering knowledge. Both theologians and scientists usually have motivations that preclude real communication. In what follows, I will demonstrate the strengths and weaknesses of both sides of their arguments.

When discussing biological subjects with religious persons, or religious topics with scientists, I often find that we are not talking about the same things. The problem of shared definition is part of the ignorance factor discussed previously. There are a couple of common misunderstandings that might be best addressed before going on to specific issues. One of them is the most commonly used phrase, adopted for everyday speech, from the theory of evolution.

Survival of the Fittest

Without conflict, there is no story. That is why there are no shows on television called "Plant Planet". Plants are uniquely lacking in bloody teeth and claws. Watching herbivores grazing on grass, or bees visiting flowers, is not very compelling except to certain fringe elements of the human race, mostly known as biologists.

Aside from predators eating prey, most animals pretty much leave each other alone. If they share the same kind of food, there are occasional, minor dustups at the source such as two bees attempting to land on the same flower. But it is never much ado and seldom lethal.

Every plant and animal that has been carefully examined has at least one other plant or animal that lives exclusively on it. Most have several and share some of these "hangers on" with other related species. The point is, that being the case, there have to be at least as many animals that live peacefully on another plant or animal, as live independently. Living together, a condition called symbiosis, is the dominant mode of existence.

Modern culture has made the concept of "fit" be about strength, speed, and competition. In fact, most of biology is about sharing space, resources, and time. No disrespect, but the so called "selfish gene" may be more naturally altruistic, helping mates build nests, watching over the young, and sharing resources for mutual benefit.

The competitive mindset is very difficult to change because of popular influences, inaccurate science, and cultural biases. But in the context of religion and science, it becomes an important distinction.

On the Origin of Life

Evolution is not a theory about the origin of life. This is greatly misunderstood even by many scientists. Evolution, as a theory, is merely applied to the existing system of life as we know it. The theory was proposed originally and specifically to account for the variability so easily observed in nature. In other words, the premise of the theory was to describe the process of change in a world that already had living creatures and plants.

Mankind has gone through its own evolution concerning the variability of living things. Genesis tells us that Adam named all the animals. But the information must have been lost because Carl Linnaeus was still trying to tackle this job in the 1700's. It appears that, in the 21st century, we still haven't finished the task!

Linnaeus invented the modern system of binomial nomenclature and named many common animals and plants of his place and time. But there are estimated to be well over a million plant species and another million plus animal species in the world. The theories that might account for differences

105

between species have been numerous. Humans have variously supposed that: God created each one originally, God has continuously created new ones along the way, and God didn't create any of them.

Deciding the ultimate cause of an event such as the "creation of life", or the "creation of an individual", is very tricky business that I discussed earlier in this book. The same problems exist in questions surrounding the origin of life. Even if we could invent a plausible story that explains life arising from materials alone, stories are not proofs. It must also be considered that God may have at least "caused" the materials and laws that created life to come together.

In reality, biology may have little to say about religious topics regardless of the popular conception. Those who understand science, but not religion, may see conflict based upon incomplete understanding. Equally, there may be some who do not understand science but are most knowledgeable on religious topics. These see conflict based upon their misunderstandings of science.

You may be surprised at how few points there are to fight about. There is plenty of room for alternative hypotheses that would allow religion and science to have a more civil dialogue.

Chapter 10 – Creation

I don't know . . .

I don't know how to create things with my hands. I mean, I have ideas that might work. But when I try to build something, it doesn't turn out. I don't know how to draw, paint, sculpt, knit, throw pottery, do carpentry, or any of that kind of stuff. I am a true product of modern education.

Creativity is a fascinating subject. No one knows quite what it is, but we all feel like we know it when we see it. The way Shakespeare strings together words and ideas has fascinated people for a long time. The huge amount of music, all created from just twelve tones in the western music tradition, is amazing. How those twelve tones can capture emotions, events and even scenes is remarkable. How a visual artist can capture an image or an idea with line, shadow and color is miraculous.

Probably few people see science as a creative endeavor. That is because they don't understand science. Do you realize that, when we describe an atom with its associated parts such as protons, neutrons, electrons, quarks and what-have-yous', we are describing structures that have never been seen? The atom is purely an imaginary construction based upon indirect evidence. Scientists can formulate an imaginary idea and then attempt some action upon it as if it were real. If what we try to do works, we conclude that what we have imagined is, at least in some part, real.

When a physician measures your blood pressure and finds it high, the physician makes an assumption based upon unseen forces that either you have too much blood or the container, your blood vessels, are too small. The physician can't see these at all but is able to draw conclusions based upon imagination.

Can imagination mislead us? Of course! We can imagine things that do not exist. We can imagine things that could not exist according to the laws of the universe. In 1958, a B-grade, horror movie called "The Blob" was released. It was about a giant amoeba that could eat people alive. There are physical reasons why an amoeba cannot achieve massive size. The film maker, however, was able to imagine it and make it appear somewhat real.

Can we imagine things incorrectly? Sure. I can imagine that the ambulance siren I hear is for my wife who just left the house a few minutes ago to go to the store. It most likely is not. Can scientists imagine a structure, or a sequence of events, incorrectly? It happens all the time, and that is why our first explanations are often called hypotheses. Until we have some evidence based upon the belief, our suppositions are just imaginary. But if we can make predictions and control events based upon the theory, then that information might someday be called a law.

I don't know how life was created. Science has theories about the origins of life, and so does religion. It might be interesting to consider the points on which they agree.

Scientific Theories of Creation

There is no one scientific theory of the creation of life. There have been a series of proposals, but all have tremendous loopholes. Instead of trying to write about them chronologically, I will try to present them as an amalgam of concepts which represent the various scientific thought.

Time: In the first place, all scientific theories about the creation of life assume the process evolved over extremely long periods of time. One of the reasons some scientists don't like the idea of a beginning to the universe is it puts a limit on the time available for life to evolve.

But the basic theory is that earth's early atmosphere was rich in some elements, poor in others, that there was a great deal of energy available, and that these elements randomly formed bonds to create the first living thing. For such random events to have occurred, it would take extremely long periods of time to accommodate the odds.

Further, even after life is created, science assumes the continuation of the process of evolution. Once again, long periods of time are required for random combinations to be selected from all possibilities. I will discuss this in detail later.

Nucleic Acids and Proteins: For a while, it was proposed that an environment, as delineated above, could give rise to nucleic acids or amino acids, the precursors of proteins which are fundamental to living things as we know them. The odds of this occurring are astronomical, of course, but it is even more improbable than scientists usually admit because the two compounds must work simultaneously. Nucleic acids are required to make proteins, but proteins are required to make nucleic acids.

So the odds of these developing simultaneously are ridiculously large. In fact, they are so large that there isn't enough time in the developing universe for this to have occurred.

Membranes: A lesser known, but probably more meaningful theory, involves the creation of membranes. The early chemical-soup of material could have formed precipitates in the form of thin films. These films could have changed the environment in two significant ways.

First, they might have set certain molecules and elements into a fixed position. For chemical reactions to occur, the two atoms or compounds to be combined must meet with enough force and be oriented properly so that they can fit together. These two conditions work together to form chemical bonds. The better the fit, the less force is needed for bonding. The larger the force, the more easily it may overcome a poor fit.

Holding one atom or molecule in place creates a situation where certain chemical reactions are favored over others. This, too, drastically reduces the odds of such events happening and can even lead to the creation of yet other molecules or films that lead to some kind of cell-membrane type of structure.

The second way a precipitate, or an artificial membrane, alters the environment is by being porous. If there are tiny gaps in the semi-membrane, then some things can pass through the membrane to the other side and others cannot because they are too large. Under high temperatures, molecules all move about randomly. Thus the smaller particles would be able to pass through the membranes and be available on both sides. The larger molecules would be trapped on one side of the membrane changing the nature of the solution on both sides of the membrane.

Once again, this situation favors certain chemical reactions over others and could lead to the formation of other organic molecules. These two attributes of membranes would significantly reduce the odds of more complex organic compounds being formed within the given time frames.

Evolution of Life: It must be noted that the original theory of evolution did not address the origin of life but dealt only with how existing species might be changed so as to create a new form of animal. Applying such thinking to the actual origin of life was a development that took place later, about the time that science took a decidedly atheistic turn in thinking. In other words, evolution as originally conceived, is not about the origin of life but is a theory that works only on existing biological systems.

There is no evidence that evolution applies to strictly physical systems of matter. While rocks change, they do not evolve in the same way as living systems. Those who propose that life originated through evolution use this idea as a basic premise with no evidence whatsoever.

So while many people like to explain how life came into existence by way of evolution, they actually have no data that can support that theory. Hence religious theories of how life originated carry as much weight as so-called scientific theories. The discussion changes considerably, though, by merely defining evolution.

Sequence: It is generally assumed by scientists that plant life had to have come into existence prior to animal life. This is because it was thought that early earth's atmosphere was low in oxygen and high in carbon dioxide. It was also rich in energy from the sun and had copious amounts of water. Granting that science's speculations about the earth's original atmosphere are correct, it is reasonable to suppose that plants developed first.

Whether plants were created the day before animals, or several million years before them, is a completely different discussion. But plants supply the oxygen created from sunlight, water, and carbon dioxide that animals utilize for energy metabolism.

Sequence is such an important concept that it must be explored further before discussing similarities between scientific and Biblical ideas of creation.

My wife and I raised four children. We learned early on that involving ourselves in specific disputes was frustrating. Yes, one child struck another; but with exploration, we always found that it was done in response to something else which, in turn, was in response to something else. There is always a previous cause. We quickly learned not to delve into the details but to deal with principles.

On the rare occasions when we had to be involved in determining what was true or not, it was necessary to have each person tell his or her story from the beginning through a sequential series of events to the end. We often found that one person would leave out certain details that another would supply. Sometimes it was difficult to determine the exact sequence, but sequence is especially critical because there is always a previous cause.

Almost any child can tell a story that sounds reasonable and sequential, except for one or two key elements. They usually don't want those elements to be known. If stories are not told in order, or if details are left out, one might conclude that the story is not true. Again, there are always previous causes.

Imagine reading a detective story to find that the book pages had been cut out, shuffled randomly, and then put back together. Even if you realigned the writing and transitions, you would still not have a book that flowed logically. The murderer might well end up being revealed before the desired time.

If order and sequence, starting from the beginning and ending at the end, is necessary for truth, then falsehood can often be assumed when details are missing, when the witness starts or stops at different places, and when things are presented out of sequence. If people reach conclusions prematurely, start in the middle with their explanations, or focus on certain details and ignore others, one can usually not expect the truth. There are always previous causes.

If a person starts back down the chain of events in an effort to determine truth, where does one stop? Modern physics acknowledges this question in something called Chaos theory. This is a sophisticated method of explaining that every event is connected in an impenetrably-complex relationship between cause and effect.

There is a story told of a man who was stung by a scorpion. He went to the river to wash out the poison. While at the river's edge, he heard a loud splash and a cry for help. He saw that a small girl had fallen into the water and was unable to swim. He rushed into the water and saved her life. Later the father of the girl thanked him for saving his daughter's life. The man replied that it was not he, but the scorpion, that saved the girl. Had he not been stung, he would not have been at the river's edge.

A famous illustration from Chaos theory is that of a butterfly flapping its wings in China. The butterfly disturbs the air and dust in such a way that changes are created in the atmosphere. Eventually, those changes create a tornado that touches down in Kansas. There are always previous causes.

If I were to twist a piece of wire into a spiral, you would recognize the order. If I twisted the spiraling wire into a second spiral, you might still see the order. If I continued twisting spirals into spirals, at some point the organization would not be visible. At this point, we would call the mass of wire "chaotic".

However, there would still be an order there. We simply would not have the ability to perceive it. The wire was twisted in a definite series of events; but because we were not present when it happened, we would not be able to perceive the pattern.

Place: It is also a typical, central theme of science that life began in water. This idea is based, again, on our theories about what the earth must have been like, the supposed salinity of early seas, and the uniformity of salinity in the cells of most living animals today. This is not a wild supposition, and it can even be considered likely.

Biblical Account of Creation

In Part I of this book, I discussed the creation. To summarize, it was unnecessary for the original writers of the Old Testament to even talk about the creation of the universe. No other religion does. But as the Old Testament was never meant to be a scientific treatise, the description provided there is in broad sequence, not methodology.

The universe was created through a flash of light which could certainly be interpreted as an explosion. I also pointed out that the Hebrew word for day translated into the King James Version of the Bible does not mean "day", but simply some long time period.

Similarly, it wasn't necessary for the early writers of Genesis to be specific as to the sequence and order of events.   Yet they proceeded to lay down a sequence of events that is amazingly similar to what science proposes. Let's follow along.

Genesis 1: 6 - "And God said let there be a firmament in the midst of the waters, and let it divide the waters from the waters." I've never completely understood the word firmament. I know it is defined as the dome that separates the earth from the Heavens. But I suspect it was defined a long time ago before our present understanding came to be. There is no real dome between the earth and sky. However, a membrane would certainly fit the description of a precipitate, especially since the waters were separated by whatever a firmament is. Speculation? Sure. But no more outrageous than other speculations that have been advanced.

Genesis 1: 9 - "And God said let the waters under the heaven be gathered together into one place . . . and the gathering together of the waters called he seas." Creating the oceans and seas prior to creating life would certainly be necessary if life were to originate in them.

Genesis 1: 11 - ". . . Let the earth bring forth grass, and herb yielding seed, and the fruit tree yielding fruit after his kind, . . ." It's interesting that the Bible claims plant life to be the first life in this sequence just as science has hypothesized. Notice, too, that the Bible specifically points out seed and fruit. Which of these might have been created first? If seed, they would have grown into plants only after there was a sun.

Genesis 1: 14 - "And God said, let there be lights in the firmament of the Heaven to divide the day from the night. . . the greater light to rule the day . . . " Now the seeds can grow into photosynthetic plants. Knowing this probably doesn't solve the problem of which came first, the chicken or the egg, the seed or the plant. It does suggest, though, that perhaps seeds were created before plants.

Genesis 1:20 - "Let the waters bring forth abundantly the moving creature that hath life . . . " Isn't it strange that Genesis specifies that animal life was first created in water, exactly like the conclusions reached by modern science?

Genesis 1:24 - "Let the earth bring forth the living creature after his kind, cattle and creeping thing, and beast of the earth after his kind . . . " Science also speculates that terrestrial life was the last to be created, and it suggests that insects were perhaps some of the first terrestrial animals.

Genesis 1:26 - "And God said, let us make man in our image . . . " Science even places man as one of the last creations in the sequence. There is a lot of disagreement about what the first man was and when he or she occurred. There appear to have been several man-like creatures. Still, man appears to be the last of Gods creations.

Genesis provides a good sequence for a story that is over six thousand years old. Keep in mind the writers had none of the evidence or understanding of modern science. They certainly didn't have to be that specific. No other religion attempts to explain not only the creation of the universe, but the correct sequence of events, in the creation of the earth. How could the writers have known this information unless the beginning was either witnessed, or the information in Genesis came from some other source than science?

I don't know the answer to that question. Possible explanations or hypotheses might include:

- the Bible must be wrong,
- or there is no God,
- or that science is correct because it provides more details,
- or perhaps because I prefer one story over the other for reasons that have nothing to do with logic or reason.

The significance of this sequence does not prove that there is a God, that science is correct, or that the Bible is literal. It does demonstrate a remarkable similarity between two supposedly opposite approaches to our self-knowledge. Again, it shows that there may be more than one explanation, or that the truth may be some amalgam of different explanations.

For example, we could hypothesize that Genesis was not written as a scientific treatise, but it actually captures some form of the truth. In that case, the claim that there is a creator would seem somewhat plausible, and science would seem to have confirmed the sequence. Thinking this would certainly not only strengthen my faith but perhaps tell me something about the power and understanding of God. In fact, it would even give me a glimpse into Gods mind and how God thinks.

I might conclude that the creation took a very long period of time and that these time periods might overlap to a considerable degree. In fact, I have scoured the Bible pretty carefully and can find no evidence or statement that the creation was ever finished. God often stated that His most recent creation was "good". And of course, after His sixth creative period, He apparently did take a break. But I don't believe it ever says that He was through creating.

In Genesis 2 He says that the Heavens and earth were finished, but it is less clear about the 'host of them". It also says He ended his work on the seventh day and rested. So what did He do on Monday morning?

Then what did he mean by "good"? Theologians generally take that as a pronouncement of either morality or a blessing. However, could God have simply meant, "There, that's good enough for now." You know, like we might say, "There, that's good enough for government work. I'll fix it later." In fact, God does not even proclaim man as "good".

In Genesis 1:26 through 30, He creates man and instructs him about the earth. Yet He never says, as He does for all of His other work, that "It was good". He does survey the work as a whole and say it was all "very good", but He does not ever say that specifically about man.

Does that mean that man wasn't good? Or does it mean that God wasn't through making man yet? Maybe God meant that it remained to be seen whether man would be good or not. Man is the only living organism who has a moral choice

Some may question that statement about choice, but it is true. Because, as far as we know, we are the only species to believe in a God. Moral choice is not about whether we choose to do something good or bad. It is about whose opinion of good and bad we accept.

If a man says something is bad, other men can always ask "Who said so?" and "What makes you the boss?" This leads to a relativistic morality, which is no morality, because everything is up to the individual. Only God can proclaim what is good or bad without equivocation. The only morality that is true morality is theological morality.

Please don't misunderstand what I am saying. I am not a scientist trying to be a theologian. I don't know whether any of my alternate hypotheses are true or not.

The point is that there are alternate hypotheses. There are many other ways of interpreting the Bible and scientific evidence. I am suggesting that conclusions drawn by some scientists about the falsity of religion and the non-existence of God, or that the claim of some theologians concerning the interpretation of scriptures, are not necessarily the only way it must be.

If there are more possible explanations, then the most we can conclude is that we don't yet know exactly how the earth was created or how life came to be. We could just as easily conclude that we need to know something more about God before concluding that there is no God.

Chapter 11 – Relatives

I don't know . . .

    I don't know why I have a hyper-mobile thumb. A hyper-mobile thumb is a little hard to describe. I can sort of throw my thumb out of joint a little, extending the second joint inward towards the palm of my hand. This makes my thumb look odd. Then if I try to move my thumb towards the fingers and away again, there is a curious sudden movement. If you listen closely, the joint makes a popping sound. My guess is that if you can do this, you know exactly what I mean. But if you can't, you don't have a clue as to what I am talking about.

    I discovered my hyper-mobile thumb when I was in about the third grade. We had these desks at school with lids that lifted up revealing metal bowls beneath in which we were to keep our books and papers. This metal bowl created a semi-drum-like object with considerable resonance. I discovered that if I rested my elbow on the desk top and wiggled my hyper mobile thumb back and forth, the desk bowl would capture the popping sound of my thumb and amplify it considerably.

    This grossed out all the girls in the room and was important to me at the time. Unfortunately, my teacher was also a girl, so she was also upset by this action. Now most adults know that little boys would rather be in trouble than be ignored. So I was repeatedly in trouble over my thumb. Oh, and girls, big boys aren't a whole lot different.

    Later I learned that a hyper-mobile thumb is caused by a single-gene, dominant characteristic.

That means that I have a gene that manufactures a specific protein that allows me to have this exalted position in the biological scheme of things. If you don't have the gene, you can't make the protein, and you won't have a hyper mobile thumb.

Unfortunately, I don't know how I got this gene or protein. I never thought to ask my Mother or Father if they had a hyper-mobile thumb or not. Now they have both died, and I can't find out. I don't really know which of my two family lines had this impressive characteristic to bequeath to me.

You Look Just Like Your Dad

How do we know people are related? Usually we can tell a lot by looking at them. I have photographs of someone in my Mother's family, at least four generations back, that has the same droopy eyes I get when I am tired. My wife is less-than thrilled to find herself looking at her husband and seeing her father-in-law. How many people have told me, "You look just like your Dad"!

How is it that family members resemble each other? Well, the answer to that question is because we come from the same cells. A cell from the mother fuses with a cell from the father, and the two become a new individual with half the cells of each parent. Everyone knows this.

The chemical molecule in the cell that carries the information that determines which proteins we will make, and what kind of eyes we will have, is called by a long, multi-syllabic chemical named deoxyribonucleic acid. Now no one wants to have to say that word any more often than necessary. You could sprain your tongue, dislocate your frenulum, or something! So we have shortened it to DNA. This is a chemical molecule that is made up of several smaller molecules that have specific shapes. These, in essence, act a little like a binary code, or Morse code, to send messages about which proteins you should make, when to make them, and so forth.

What is interesting, in the context of this chapter, is that every living thing uses DNA to carry this information. This means that chemically, you are a lot like a philodendron, a termite, or even an amoeba.   When we examine living things, we find that we are all quite a bit alike. We all use DNA to make new cells and create new life. In fact, scientists estimate that about fifty percent of our DNA is similar to that of almost all other living things. In some cases, the percentage of sameness is as high as ninety-nine percent. So if you go out and hug a tree, you are hugging a cousin. Okay, a pretty distant cousin. Not one you would want to marry I suppose.

That isn't all. You have another molecule in you that is universal in living things. It's called ATP. If you really want to know, that stands for adenosine triphosphate. This molecule is used in energy metabolism of both plants and animals. There are many other proteins that are universal throughout the plant and animal Kingdoms, as well as others that are nearly so. I just read where it's been discovered that bees have an insulin-dependent pathway. This pathway is involved in determining whether a bee will become a queen or a worker bee.

In other words, it seems that we all belong to the same "family tree of life". Living things are all made of the same stuff, and chemically, living things look pretty similar. Understand that this relationship is not necessarily linear. That is, it isn't like the biblical language where an amoeba begat a philodendron, which begat a tree, which begat a grasshopper, which begat . . . You get the picture. In fact, maybe nothing begat anything else in particular. Maybe we are all just cut from the same cloth.

Some people get upset about the thought that they might be related to monkeys. The truth is that it is even more strange than that. Life seems to be one, single family tree, all sharing DNA, ATP and many other chemicals. Humans, as we are now, are not related to monkeys. We probably are not even descended from them. They have DNA, however, that is very similar to ours. So does a philodendron.

One claim of biology is that living things represent a single pattern of some kind that is based upon a similar diagram and blueprint. Should that be so surprising? This is not an argument for evolution. In fact, I see it as greater evidence that this pattern of living things, along with the natural laws of the material world, originated from something other than chance. Have scientists speculated on ways life could have started by chance? Sure. Have they proven it? No. And as pointed out elsewhere, stories are not proofs. Unexplained patterns, laws, and relationships, though, are certainly better evidence than speculation.

Individuals

Most of us know, by way of the media, that DNA can be used to determine paternity. It can even be used to determine guilt or innocence of certain crimes. While all living things are made of DNA, each has its own unique arrangements. Despite the fact that our family likes to say how much alike my son and I are, especially when either of us has performed in some particularly negative way, we both know that there are also some significant differences. So in spite of the fact that we all seem to be cut from the same cloth, we all seem to be unique as well.

Variation is so obviously a characteristic of our biological world that it is hard to refute. It may seem like all bees look alike. But those who study bees see a tremendous amount of variation from bee to bee as they study them closely and learn what their distinguishing characteristics are. In fact, biologists now believe there are something like 7000 species of bees in the United State alone.

The number and kinds of plants and animals in the world is truly astounding. We have not even come close to discovering, describing, and naming them all. Some kinds of plants and animals have disappeared from the earth in my life time. Others have died out during the period of recorded history. Many became extinct even before we knew they existed.

Darwin's original attempt was simply to try to explain how such a variety of plants and animals came into existence. He did not address the origin of life, only the source of variation. That source of variation now appears to be DNA, and it is shared by all life forms. This knowledge proves nothing about origins or God any more than knowing the fact that all sky scrapers contain steel explains anything about the origin of skyscrapers or the architects who did the plans for them.

The fossil record indicates that there once were many plants and animals on the earth that no longer exist. We know that at least some of those ancient animals also shared the DNA code. What is most interesting about fossils, to me, is that they indicate what conditions may have been like in the past. Studying fossils is about the nearest thing we have to a time machine. The fact that ancient life used the same chemical language as modern life seems only reasonable.

Shared Genes

Does sharing a chemical language with plants and animals demean humanity? Does the ubiquity of DNA indicate a random universe, or would it be most reasonable to think it indicates some kind of design? Design is forbidden by some scientists but often for reasons that have nothing to do with reason, logic, evidence, or science.

On the other hand, many religious believers think that a shared chemical language somehow demeans them, or demeans the actions or intent of deity. They do not offer any explanation of why sharing such a commonality with God's other creatures would be demeaning.

Evolution does contain an anti-religious theme, especially in how it is presently taught and interpreted. I believe that is why most Americans remain skeptical of it. The key towards understanding is in "how it is interpreted and taught" today. Those who teach evolution are seldom knowledgeable about religion, and those who mistrust science are often not familiar with science. It is generally thought that the Bible contradicts evolution or the other way around.

So just what does the Bible say?

In Genesis 2:7 it reads, "the Lord God formed man from the dust of the ground and breathed into his nostrils the breath of life." The earth may have been formed from nothing, or from firmament, or something else. But man was formed from a specific thing, the dust of the ground. Man was formed from the elements. Since all other living things were created after the earth was created, but before man, this in no way means that they weren't created from the same substances. It is not inconsistent to say that man is made up of atoms and molecules like every other living thing.

Where chemistry is concerned, man is definitely similar to all other living things. This is not demeaning, nor is it evidence that this similarity is random. In fact, the simplest and most straightforward interpretation of these similarities is that chemistry was part of a plan. Yet many want to make an issue of the fact that we bear a chemical, and sometimes physical, resemblance to lower animals such as the apes. Should not organisms that share DNA code and protein structure not share some physical and physiological characteristics?

Yet the majority of people continue to reject this interpretation, not on religious grounds, but because we intuitively sense a significant difference between apes, or any other animals for that matter, and humans. Of course, science does not claim that we descended from apes, only that apes and humans descended from a common ancestor. That is, we share some physical traits because we are composed of similar chemicals. Well, like I said, we share chemicals with philodendrons too.

Look again at the quote from Genesis. It states that man is different because God breathed something into man. Christians, and many others, believe it was man's immortal soul. No reference to such an act by God is made in conjunction with the other creations.

There really is no theological problem with assuming that our physical bodies are similar to those of other creations. The difference comes in the belief that we, ourselves, are unique and created in the image of God. Perhaps the difference is the very thing that has been breathed into humans but not into other animals.

Behavior

There are actually many differences between humans and other animal life. Scientists acknowledge that there are some physically-unique characteristics like opposable thumbs and walking upright.

However, many people dismiss other major differences as being cultural, as if culture is common to all living things. In fact, no other animal besides the human has "culture". I don't mean there aren't rules of the pack, but unique behavior and attributes set humans distinctly apart from other animals.

For example, sitting. No other animal sits the way humans do. More specifically, no other animal sits down at a table to eat or has prescribed times for meals. Further examples might be:

- other animals simply eat whenever or wherever they want,
- other animals do not build and decorate homes in the same way humans do,
- other animals don't wear clothing, or decorate themselves,
- no animals, as far as we can tell, perform collective rituals such as blessings on their food much less religious rites of any kind.

Even if you don't believe in religious rites, it must be admitted these are unique characteristics of humans. They demonstrate a different kind of thinking and existence than is found in other animals. Perhaps religious belief is not a primitive superstition but an advancement over animal existence.

Of course, the Bible is not proof of one thing or the other. But then, science has not proven that there is no soul either. Souls, spirits, or whatever you might call the abstract essence of a human, being abstract or non-material, are simply not anything science can investigate. I present the above ideas concerning the Bible, as I have elsewhere, to show that there are other interpretations besides the atheistic interpretations favored by some.

We share chemical makeup and physical form with other animals. Is that contrary in any way to Christian belief? Or is it a source of inspiration and

joy? Sharing is a central tenet of Christian theology.
As Paul taught in his letter to the Corinthians (I Cor.
10:17), "For we being many are one bread, and one
body: for we are all partakers of that one bread."
Again in I Cor. 12:13 Paul says, "For by one Spirit we
are all baptized into one body, whether we be Jews or
Gentiles, whether we be bond or free; and have been
all made to drink into one spirit."

In these chapters, he expounds on how our
diverse community can be as one. Each part can be
different from each other part, yet something
essential to all the parts runs through it. This
sameness has been a universal theme among
Christians for centuries.

Space and Place

What people often do not realize is that the
sharing of space and place is also an essential theme
of biology. Many biologists and other scientists even
ignore this universal theme, a theme that is probably
more foundational than the concepts of evolution,
change and creation. Darwin emphasized
procreation and competition which can be related to
space, but he did not address these topics directly.

Some things are so fundamental that humans
tend to overlook them or even take them into account.
I think that is what has happened to the idea of space
and place in the sciences, especially biology.

Life is a surface phenomenon. Living things
don't live "free-floating" in the air. Birds may fly
through the air, but they land and live on the ground
or in trees. Fish may swim through the water, but
water is held to the earth by surface tension. Too,

130

fish either return to the shallows to reproduce, or they use the surface of the female to nurture their young.

This is another of those obvious things. We tend to ignore its significance except in specific instances. For example, scientists fully recognize that the cell membrane is an action interface with the outside milieu. Biologists have mapped out how the surface of DNA must literally "fit" with the surface of the opposite strand of DNA molecule.

There are other instances where we recognize the significance of place to living things. Imagine for a moment that the earth is a smooth sphere. The limiting factor for all of life would be the available surface area. Once the smooth surface was covered with life, there would be no place to go for new life to grow.

Of course, the earth isn't a perfectly smooth sphere. It has mountains and valleys. The only way to increase surface area on a smooth sphere is to either buckle the surface outward, or collapse the surface inward. In fact, that is how the earth is constructed with mountain ranges and valleys.

And there are other restrictions to consider as to where life can exist such as water availability, sun exposure, the size of the sphere and so forth. One of the limiting factors for all life forms, however, is adequate and inviting surfaces. I will discuss this concept further in a later chapter. For now, when surface area is limited, the only "place" for other living things to grow is on other living things.

That is why it is so common for living things to live on other living things. What greater surface than a surface that shares your chemical makeup, has a behavior that is favorable to living things, has plenty of water, and is shaped by the same chemical codes? Humans in the 21st century have been conditioned to think of things living on them as bad. More often than not, they are benign and, occasionally, even beneficial. Indeed, we are all "being many, one bread."

Stewardship

When I first began the study of biology, I had friends who gently questioned me about how I could go into a field that promoted evolution. So successful have certain scientists become that biology has been reduced in the eyes of many to the study of evolution. As such, biology is sometimes discarded as an unfriendly field of study.

Biology is the study of life, however, and life is one of the sacred attributes of most religions, especially Christianity. How could it be an improper study for man? In the first chapter of Genesis, God said to Adam and Eve, "Be fruitful, and multiply, and replenish the earth, and subdue it: and have dominion over the fish of the sea, and over the fowls of the air, and over every living thing that moveth upon the earth."

To replenish is to restore plenty. Dominion is sometimes read as if it means domination, but that is not its meaning. To hold dominion is to hold the power of life and death over something. Can anyone deny that man holds dominion over the earth? Perhaps one of the most important things man can study is how to replenish, and avoid the death of, the earth.

Furthermore, in the second chapter of Genesis man is told to "dress and keep" the garden and name "every beast of the field." In these things man was given stewardship over the earth. Stewardship is a word that suggest wise and prudent study and conservation of the place we've been given to live. Man has sometimes failed at this assignment. Sometimes it's been through greed but, more often, through ignorance.

Chapter 12 - CHANGES

I don't know . . .

I don't know how to tell a dog from a cat. Well, I guess I can, but I can't tell you how to do it. What would I say? A dog has fur? So does a cat. Dogs have tails? So do cats? Dogs have canine teeth, and so do cats. How could you explain the difference between dogs and cats in a way that is easily understood by all? Yet toddlers learn to make this distinction very early in life.

Here is another thing I don't know. How do you tell whether or not someone understands a concept? It doesn't matter whether it is a set of instructions or the concept of dogs and cats. In the case of the toddler, we would never expect the child to explain the difference between dogs and cats. Adults can't even do that. No, we would show them a lot of examples. Then when they could routinely, visibly distinguish one from another, we would decide that they understand.

How do we know if someone understands our instructions or not? Do we quiz them? I think we usually watch to see if they carry them out accurately. How do we know if someone understands an explanation? What does that person look like? What does a person who understands look like compared to a person who does not understand? Since we all want to appear to be in control, most people quickly master the ability to look like they understand, whether they really do or not.

I ask these questions because understanding and knowing are often much more difficult than humans understand or know. We often think we know more than we do. This flaw plagues theologians, scientists, and even regular people like me. The rest of this chapter is a discussion about what we don't understand but often act as if we do.

"After His Kind . . . "

What "kind" of animal is a dog or cat? Biologists call them mammals. But there are a lot of different kinds of mammals; apes, mice, whales, and bats are a few. On the surface, they don't look much alike. Are they the same kind?

One of the concepts taught in Biology, that sometimes causes problems for religious leaders, is the idea that plants and animals can change over time. The fact that living things can, and do, change is a readily observable phenomenon. Even during the short span of a human life, change is ever present. Therefore, it is hard to take exception with the idea of change. The infant does not usually resemble the old man he will become. Of course, there could be an argument about "how much change" and "how quickly it occurs", but that living things do change over generations is pretty obvious.

Many religious leaders have decided that plants and animals cannot change over time. They all must have existed, and will always exist, in the exact same form as they do today. I do not understand where this idea comes from. The Bible is not a noted scientific text. However, as we have seen, it is often surprisingly accurate. It is the main source for understanding the Christian view of the world.

If we examine the Bible, references on the nature of living things are in the first chapter of genesis. In Genesis 1:22, after God created most of the living things, He tells them to be fruitful and multiply in the earth. It does not say that they are expected to make exact copies of themselves.

There is one other phrase in Genesis that might be concerned with the accuracy of reproduction. In Genesis 1:11-12 it reads, "And God said, Let the earth bring forth grass, the herb yielding seed, and the fruit tree yielding fruit after his kind . . . " Apparently it is the phrase "after his kind" that needs interpretation.

I am told that there is some disagreement about the translation of the word "kind", but I am not a language specialist. It is probably more instructive, for a discussion of religion and science, to discuss what a biological "kind" is, and how to name a different "kind".

Is a "kind" like a Phylum? A phylum would be whole large groups of plants or animals such as mammals, arthropods, mollusks, and so forth. In that case, reproducing after its kind would certainly include a great deal of variety. But a kind might also mean a species, or even a particular set of genes. Remember, we tend to look somewhat like our parents, but not exactly.

I tend to favor the following definition. God created cattle after their "kind" which means all forms of cattle that can interbreed with one another. There are numerous kinds of cattle. I guess God populated the earth with a variety of cattle types from which new varieties could arise. Or perhaps God made one kind of cow with the ability to produce variable offspring.

Does anyone know who the first geneticist was? As far as I can tell, it was Jacob from the Old Testament. His story is told in Genesis which means the tale, itself, is at least 6000 years old. According to the story, Jacob worked for his father-in-law and was in charge of the herds of cattle for twenty-one years. It is recorded in Genesis 30:29-41 that Jacob used corrals to control the breeding of the animals. He did this work because part of his pay was to be the spotted animals. His father-in-law wanted the solid colored animals. So over the years, he carefully bred the stronger animals with the spotted animals to increase and strengthen his own herds.

Admittedly, he was not creating a new species. However, this is an example of the fact that the Bible does not claim that "kinds" of living things can't change. In fact, humans have known that things change and that change can be manipulated over a very long time.

An additional problem becomes the naming of a kind. For example, wolves, coyotes and foxes look very much like the dog that chases a stick for you and sleeps on your couch when you aren't looking. But we have different scientific names for the two kinds. The wild one might be called *Canis lupus*, and the one in your house *Canis familiaris*. Then we have different names for different kinds of *Canis familiaris*. There are Collies, Shepherds, Chihuahuas, and Pekingese.

In summary, the Bible does not contradict two important scientific findings. All living things belong to the same group utilizing DNA, and all living things can change.

Names

Naming something is a powerful mental activity. It is such a common occurrence that we generally don't think much about it. In fact, we have names for just about anything we can pick up and hold in our hands. Come to think of it, we also have names for all kinds of things that we can't hold in our hands: things like love and freedom.
The reason why naming things is so powerful is that, as soon as we create a name for a specific item or idea, that name can become a symbol for the idea or item. First we name a physical thing, and then we can use that same name for a non-physical concept.

We can name an animal a dog, and that dog is something we can put our hands on. Then we can talk about dogs as an idea or group of organisms with similar characteristics. We can even use the word "dog" to refer to a human who, we believe, is acting like a dog.

We can give a physical thing the name of a color, and then use that color to stand for an emotion like anger. We can also give a name for an emotion, such as love. Then a tangible item, say the heart, can be used as a symbol for that idea.

Naming things is a great achievement because it not only identifies an object for future reference, but it identifies mental categories for discussion. While humans generally think that they are naming things to create individuality, they are also naming things to help us identify the common features of similar groups. We might name a child to create individuality, but the family name collects that child into a group called a family made up of related members.

Adam was asked to name all the animals. He either named all there were at the time, failed to do the same, or the record was lost because humans are still giving names to animals today. In fact, we have named over a million to date.

Scientific Kinds

One problem scientists continuously face is determining "a kind" and how it should be named. Scientists call a "kind" a "species". The definition of species, however, is nearly as open to interpretation and debate as is the religious discussion of "kind".

A species is often defined as a group of organisms capable of interbreeding and producing fertile offspring. In many cases, this works just fine. But what does one do with organisms that don't reproduce sexually? Did you know that there are even animals that can reproduce parthenogenically, which is with virgin females, and no sexual reproduction?

Then there are bighorn sheep in Colorado and in New Mexico that live several hundred miles apart. Theoretically, the two could reproduce but, practically, they never do because neither would venture into the type of habitat that separates them. Are they of different species?

Common names for plants and animals often correspond to names of a species. "Lions," "walrus," and "Camphor trees" all refer to different species.

"Deer", on the other hand, refers to thirty-four different species. As we learn more about various animals, we sometimes decide that two animals are separate species even though we used to think they belonged to the same one.

Because there are so many living things to keep track of, scientists have invented a categorical system into which they place similar plants and animals. At first we grouped them according to their anatomical similarities. But in some cases, organisms have physical similarities but distinct physiological differences.

For example, many desert plants of the world have similar characteristics such as needle-like leaves. These have developed to minimize water loss. Such plants often look alike but may use entirely different photosynthetic mechanisms to convert solar energy to sugars. In which species should they be included?

With the advent of the theory of evolution, most biologists began grouping animals together that were believed to have the same ancestors. Whether we use physical similarities, physiological similarities, or presumed relationships to previous generations, we are simply naming specific plants and animals so we can talk about their characteristics and be able to identify them.

With millions of names to keep track of, biologists next created a system of categorization to help them keep track of specie names. They called this larger category a genus. But there is a big distinction between a species and a genus. A species is defined by a specific, physical attribute even though we might argue about certain critters in specific instances. In contrast, a genus is a category, a mental invention of the human mind. It is a type, or an abstraction, not a reality.

For example, when we talk about the genus *Canis* (dog) we have created a mental category that cannot be held in one's hand. Things that cannot be held in our hands are abstractions or ideas. So while one can hold a Chihuahua in one's hand, they cannot hold all the animals that are collected under the category of *Canis* in their hand.

Scientists deal with the same questions and difficulties as theologians do in trying to come to grips with what a "kind" is. Science seems to have developed a logical and consistent method of naming things, but even then, there is often confusion and disagreement.

How Change Happens

Okay, so where do all of these different kinds of plants and animals come from? How did we get such a variety? Well, that is exactly the question that Charles Darwin asked. His original work was called "On the Origin of Species", and all he was attempting to do was explain how such tremendous diversity might have come into existence. It is clear that he did not set out to circumvent Gods role as creator. He simply saw a great variety of living things and observed that even organisms of the same species were not identical.

I am going to review the assumptions of evolution here because I wish to make a point about scientific and theological thought. Many people have summarized this theory. The explanations generally go something like this:

- individual "kinds" vary from one another,
- much of this variation is inheritable,
- most "kinds" are fertile enough to make their population increase,
- population numbers tend to remain about constant,
- environmental resources, such as food, are limiting factors for offspring survival,
- individual varieties, less suited to the environment, will not survive to have offspring,
- this process slowly selects certain variations over others and creates new "kinds".

If we recognize that humans can breed changes into living things, then we cannot help but acknowledge that nature might also select attributes. That selection might also result in change. However, this does not have to be interpreted by scientists that there is no God, or that God did not create the world as some scientists insist. It may be that God created living things with exactly this plan in mind. Recall that it never says in Genesis that God finished creating everything. It only states that he rested.

A Serious Difficulty

There is a serious difficulty with the theory of evolution that is seldom discussed. The difficulty comes with the fact that the environment is the factor that influences which plant or animal survives. However, the environment is, to a large extent, made up of the plants and animals themselves. Thus, the selection of the organisms that survive changes the environment. In turn, the new environment selects new organisms.

Evolution can be compared to a dog chasing its tail. The environment selects and changes the organism which in turn changes the environment. The process never ends, and the dog never catches its tail. This does not disprove the theory of evolution, though it does cause one to think about the significance of the study. It provides the perfect definition of those who are "always learning and never able to come to a knowledge of the truth".

Could it be that biologists discovered the great "eternal round"? Does that definition seem a little less threatening than "eternal evolution"? Could evolution be explained by recognizing God as an eternal creator, Alpha and Omega, the beginning and the end? Obviously such an interpretation would require changes in the basic assumptions of both scientist and theologian.

Conclusion

Again, I am not advocating the acceptance or rejection of evolution or theology. I only hope to point out that both camps could have grossly misinterpreted evidence and come to erroneous conclusions. Perhaps, as Tevye says in the play, "Fiddler on the Roof", they are both right.

CHAPTER 13 - Mutations

I don't know . . .

I don't know why my mother had one eye that drifted to the right. Hey, I didn't even know that she did until I was nearly twenty-two years old. Apparently it was due to some eye-muscle problem, and she didn't have it corrected until I was completely grown. The effect made it difficult to tell where she was looking when she spoke to you.

The truth is her eyes caused me a lot of distress at one point in my life. My wife and I were married in Europe while I was stationed there in the army. In the course of a typical, newly-wed discussion, my wife wondered aloud what our children might be like. She hoped they would not have trouble with their teeth like she had or trouble with their eyes like my family. I hastened to explain to her that I was the only one with very bad eyes in my family. Then she said she wasn't as worried about poor eyesight as she was about my Mother's crossed eyes.

Understand that, for most of my life, I avoided direct eye contact with my Mother, as I was more or less guilty of something all the time. So I really did not know, at the age of 21, that my Mother had an eye that would technically be called a wall-eye. It didn't cross but went off laterally to one side. Being ignorant, and having felt that my Mother had been attacked, I replied in a gentlemanly manner that my wife's Mother wore combat boots too.

An intense discussion ensued but to no avail. One could not exactly call home, half-way around the world, and ask one's mother if she indeed had crossed eyes. I had to wait fourteen months to be sent state side before I could look my Mother in the eye to see that she really did have a wall eye.

As I said, I don't know why she had that eye. No such medical situation has shown up in any of her offspring out to even the third generation. I guess it was just some kind of birth anomaly or mutation. A birth anomaly is a fancy way of saying an accident of development that we can't explain.

But a mutation? That is something different and quite specific since it has a lot to do with evolution, or change in living systems, and consequently creation and theology. Perhaps we should examine mutations more carefully.

What is a Mutation?

It turns out that a mutation isn't just one thing. There are at least two ways in which a mutation can occur, and there are sub varieties of those basic two methods. In simplest terms, a mutation is a change in a set of instructions that is inserted when a copy is made of the instructions.

A person copying instructions can switch two letters, leave a letter or word out, make a smudge so that the message is difficult to read, or get lost in where to take up after a pause. Likewise, repeating, deleting, or switching instructions within a cell can occur resulting in a new cell, or some component part, being changed.

Errors can also occur in reading the instructions, not just in making the copy. The copy might be entirely accurate. But if the instruction is misread, an error can still occur in the final product.

145

Both kinds of errors can happen, and it is often difficult to tell which type of mutation is the cause: copying or reading.

Have you ever tried to copy a quote in longhand or by printing? I have dyslexic fingers, so I have trouble even typing a lengthy quote without error on a key board. For that matter, have you ever copied something on a copy machine only to find that dirt on the glass, or smudges on the roller, gave you specks of dirt on the copies?

A mutation is a copying error. Some kind of information is being copied, and a flaw or error is introduced into that event. In the cell, information is carried in a series of molecules that are arranged in a specific order. This order makes up a chemical code. It is not strange that molecules can carry information. Many languages have different alphabets, but all say similar things. Messages can even be sent with the dots and dashes of Morse code or the ons and offs of electrical impulses in computer languages.

Occasionally, a molecule in the message is either miscopied, or misread, resulting in a new message. This is called a mutation. As you might expect, many changes are detrimental to the organism. If you change something that works, the most likely result will be that it will no longer work.

We also need to recognize that what we call copying "errors" may not be "errors" at all, but simply copying differences. My son says some people are irritating. If I were to agree with him, I would probably say they were annoying. One word isn't really "right" or "wrong". In fact, the person in question may only be "aggravating". The word "errors" implies something is wrong, when in fact, it may only be different. The difference may actually be anticipated and desired like when we carefully breed animals to our own specifications.

Then, not all mutations are bad. Let's say you are printing a photograph of a friend that was taken on an outing. The sky in the picture is blue, but there are no clouds. What if a smudge on the printer causes a dark blur to occur in the sky that looks something like a cloud? In doing so, it changes the picture making it far more dramatic than the plain blue sky. You might actually like that "mutation" and wish to perpetuate it in your photographic copies. In the same manner, mutations may sometimes, though rarely, improve a plant or animal's survivability.

The Genetic Language

I'm not sure it's necessary to go into the details of the genetic language in this book. Scientists do know quite a bit about it. It might be helpful in this context, however, to know how a change could occur. It seems reasonable to understand that a person copying something in English might write the word "right" instead of the word "write" in his haste or inattention. Or someone might use the word "personnel" instead of the word "personal". I have even written the word "walk" instead of "run" when copying a sentence because my mind was thinking ahead and anticipating the word "walk".

In somewhat the same way, genetic language can be changed. The genetic language is made from a chemical code where specific molecules occur in a specific sequence. The sequence is read in clusters of three items at a time. That is, each word in the sentence is three letters long. But there are four letters possible. Since this is understood, there is no need to put a space after each word. One word follows the other instantly: itwouldlooksomethinglikethisinenglish.

However, since the words are run together, a change in even one letter changes the groupings of all the following letters. In other words, if the sequence looked something like this: itwouldlooksomethinglikethisinenglish, the individual genetic-type words would be itw-oul-dlo-oks-ome-thi-ngl-ike-thi-gli-ket-his-ine-ngl-ish. However, the actual code has only four chemicals. Using their initials, the code would look like this: atctcagctcgt. It would be read: atc-tca-gct-cgt.

Note that if I add one letter, or delete one letter, it would change all the other groupings. For example, if I dropped the third letter "c" in the above example, the following groups would read: att-cag-ctc-gt. This is completely different from the original list of genetic words and will result in a completely different outcome. Itwoullooksomethinglikethisinenglish, but be even less intelligible.

Most mutations happen like this and create whole new sequences. Since the smooth operations of the cell require all the elements to work the same, the most likely result of a mutation is loss of function. In fact, the overwhelming majority of mutations are probably lethal to the cell in which they are produced.

Variation

Therefore, mutations do not account for a lot of the variability we see in the biological world. Sexual reproduction allows living things to introduce variation into the population with low risk of failure by combining the gene languages of two successful organisms. This type of reproduction, where two successful living creatures combine their DNA messages into a new, third organism, is the primary way that variation and individuality are introduced into the world.

In fact, as a side issue here, this becomes the definition of life. Instead of relying upon some guess as to when the spirit enters the embryo, why not adopt a definition that simply defines new life as occurring when a new DNA combination successfully divides to complete a new cell. This is a scientifically verifiable moment and is not open to argument or interpretation. Of course, it would not meet the social and political requirements of a portion of the population at present.

Much of science is concerned with physical events that happen the same way every time. Mathematics describes things that can be explained the same way each time within a group of possibilities. In physics, apples fall in the same manner as all things fall once we understand the parameters. In chemistry, we attempt to get certain reactions to occur in the same way every time we mix the proper ingredients.

In biological systems, we must deal with those events and objects that do not happen the same way each time. Variability is the law of living organisms.

If the environment changes dramatically and all the organisms remain the same, the organisms will all die. Only if there is a spectrum of possibilities, can the organism be sure of survival.

Therefore, the introduction of variation, with low risk of the variant being dangerous, is a great advantage to the survival of living things. The ideas behind the study of evolution are simply arrived at by scientists who are trying to understand the mechanisms for a given variation in order to determine the viability of that kind of variation.

What is a Law?

One part of the great debate about evolution is over whether it should be labeled a "theory" or a proven "law". Scientists like to treat evolution as an established law. In some ways this is useful, but not entirely accurate.

Too, people quibble about the meaning of the words law and theory. I'm not sure this matters a lot as long as we agree on when the word "law" applies and when it doesn't. Of course, humans seldom agree about anything. It's like waiting at a stuck traffic light at 2:30 in the morning. There is no traffic in sight, and it's been three minutes. I know what the law reads, but the label of a theory seems a better fit at the time.

In general, a law is a statement that describes what will happen given certain events. If you step off a cliff, you will fall. That is a pretty reliable law. If you run a red light, you will likely have an accident or get a ticket. These results are fairly reliable. People have survived running red lights without accidents or tickets, especially at two thirty in the morning. Still it is a description of what could happen if you make this behavior a practice.

A law usually allows us to either predict or control events. The laws of physics and chemistry are often applicable in some very specific and useful ways. For example, the various gas laws can help us predict explosions and control machinery.

Evolution can't be an actual law, then, because it only predicts change in a very general way. And while we can use the concepts for breeding to achieve some control, Kentucky Derby winners still don't always produce Kentucky Derby contenders. Predicting or controlling the environment, and what the individual organism's evolution will create in the future, is futile. In this way, evolution is probably not a law in the same way that our physical laws are.

What is God's Role?

Does God have a role in evolution? The answer to that question obviously depends on whether or not we believe in God. But while many scientists say they don't believe in God, they still accept the belief that the universe is an orderly place with rules and laws that seem to be universal in time and space. Since they have no explanation for how those rules came into existence and why they exist, might we safely assume such people are being obtuse?

God's role in the creation of life, and the explanation of variability, seems to depend on whether we choose to call him "God" or "Universal Order". Do we see God as highly involved in the details of everyday life? Or do we see Him as the power that set the stage and then lets things unfold on their own?

For example, if we decide where to eat out tonight by flipping a coin, does God have a hand in determining whether it comes up heads or tails? Or did He simply create humans so that they had the capacity to make decisions based upon coin flips?

The same thinking applies to any role God might have in evolution. If we see Him as a hands-on God, then He can control which mutations occur and which plants or animals are successful at reproducing and which are not. If, however, we see Him as a hands-off God, then He doesn't involve Himself in such matters. He simply designed the system in which change occurs.

The biologists view is that we have no idea how a coin toss would turn out. All we can do is determine the probability of the coin toss being heads or tails. It is the same with mutations. We don't know when and how they will occur, only what the odds are that they will occur. Whether or not God steers breeding in a given direction depends on what we mean by direction.

Chapter 14 – Direction

I don't know . . .

   I don't know if I know what right and left actually mean. I know which sides are my right and left, but these terms have some peculiar limitations. There are unequivocal directions, of course. They are forward, backward, up, and down.

   I don't always know what the word "direction" means when applied to abstract things like love or democracy. Why do we fall in and out of love? In fact, why do we even fall? Why don't we jump in love? Maybe we should jump up to love and then fall out of it. What if we fall in love and can't get out? That's what happened to me. But how can we "fall" in both directions?

   Some events can be confusing. I understand sunrise and sunset. But do they show progress? I mean, do they have a direction? No. The change from one to the other happens in a cycle, and my sunset is someone else's sunrise. Even if I restrict my thoughts to my experience, one isn't exactly better than the other. I have seen a lot of sunsets that I was perfectly happy to have happen, even overjoyed. Some sunrises are exciting, but I have dreaded a few as well. Simply, they occur in a cycle.

   I think everyone understands cycles. Most of our lives are circular; in sunrises and sunsets, lunar cycles, and annual cycles. Yet humans continually expect time to travel in a straight line as if it is leading some place. Even if there is a part of time that is directional, there is no guarantee that the direction is up. Straight lines can go down as well.

One of the debates about evolution is about whether or not it has a direction. Religious folks like to think in terms of improvement and becoming better. Scientists like to think in terms of randomness. Neither group really thinks in cycles. Even among scientists, there is a debate about whether or not evolution has a direction. I have never heard anyone talk about evolution being cyclical.

I find all of this somewhat strange, since we measure time by cycles. But humans never seem to see history or evolution as cyclical. Since this concept causes some confusion and argument, I thought I should examine it a little further.

What Does Direction Mean?

Actually, I don't know that either. I think, a lot of the time when people argue about direction, they aren't talking about the same thing. There are seven different definitions for the word in Webster, and that doesn't count the nuances within those definitions.

One of the meanings of the word "direction" is "guidance or instructions given concerning conduct". Many scientists don't want a guiding set of instructions for life because it would imply a God and, for various reasons, some scientists don't want God in their explanations. Sometimes there is a sincere desire to learn how God does what He does. That may seem cheeky to some and ultimately curious to others.

The word "direction" also means a specific instruction, or an order. This is slightly more acceptable to many because it implies that God ordered changes to happen but doesn't have a direct hand in it all along.

Some scientists want to believe that life evolved from non-living material, but that requires that the laws of evolution also apply to physical objects. This has never been shown to happen. At least an equal number of folks want God involved every step of the way.

"Direction" can also mean a "channel" or a specific course to be traversed. This definition implies that there is a destination to be reached and a single best course for getting there. There are those who believe mankind is on such a course. These people believe that mankind's progress is toward something superior. Examining the path mankind has taken does not lend much evidence. Individual lives may improve over time, but that seems to be peculiarly disconnected to generations. Biologists run into another problem in that they usually cannot define the destination.

"Directing", a similar concept, is defined as "being the prime motivator and organizer". A "director" provides direction to an orchestra, a play, a business, or other types of organized efforts. Directing implies a coordinated effort on the part of many separate entities. One may exist in the form of God, but scientists have not been able to discern His presence. However, the God theory could be true, and scientists still might not know because:

- they aren't educated in theology,
- His ways are not our ways, and His thoughts are not our thoughts,

- He doesn't want to be concretely known for any number of reasons that hinge on faith.

The point of all that is to simply show that when we start arguing about whether evolution has a direction or not, we immediately have to agree on what we are talking about. We must then hold the agreed upon definition firmly in mind. Humans seldom agree. Even less frequently do they hold anything firmly in mind.

It's Time to Agree That We Disagree

It's difficult to discuss direction without discussing time. It would be difficult to tell which direction a circle is facing. Yes, a non-uniform object can be said to face in a certain direction. But usually, we tell direction by changes that take place in space and time.

One of the arguments concerning evolution and direction evolves around the time necessary for changes to take place. Some fundamental, religious folks want all living things time frame to be only six thousand years. Scientists generally see life going back much farther, perhaps even millions of years. The evidence seems pretty clear, though, that physical and biological history is older than six thousand years.

It is intriguing that alphabets, written languages and number systems, all seem to date back to about six thousand years ago. Perhaps there is room for interpretation there. Perhaps humans are a special creation, with special abilities, that dates back six thousand years. A possible explanation is that human-like creatures that lived before that time may have been a different "kind".

While we might make assumptions about the total time it took for God to finish the creation, assuming he is done, there is another element of direction over time that is generally ignored. To repeat myself, modern man tends to see time in a straight line while the world revolves in cycles. Life itself is also cyclical, and we measure evolution according to generations.

Consider that not all generations might not be of equal length, and so not all generations exist under the same environmental-selection attributes. Plants tend to complete life cycles within the seasons, but some plants are annuals, completing a generation each year, while others are perennials which means they are exposed to different conditions over a much longer period of time.

Some insects' complete multiple life cycles within a season which means the environment of one season may exert tremendous selection pressure on the population. However, the next year's season can be radically different and exert completely different selection pressures. Long-lived animals such as humans and whales may face very different selection pressures that would change from season to season and year to year. Since a human may not be reproductive for 12-24 years, there is a tremendous variation in potential environmental selection pressures.

I believe that these differences in life-cycle periods are generally ignored by biologists. However, the direction of an organism's development through generations must be impacted by the length of the organism's generation.

Agreement
What if we accept that, by "direction", we mean some kind of improvement in the organism? Now we have to define what an improvement is. Is something that becomes more complicated over time an improvement? Or should we define improvement as an increase in success? How should we measure living success? Is it through an increase in numbers, biomass, length of life, or yet another parameter?

What if an organism becomes more complicated but has a decrease in numbers? Mammals are not as common as tapeworms. Or what if something gets less complicated but increases in frequency? Tapeworms have fewer systems and are relatively simple compared to mammals, but there are more of them.

Infirm Minds
Well, all that is giving me a headache. Let's just assume we can arrive at a definition of "improvement". That still does not inform us as to what has changed. Biologists generally study plants and animals, but it is the environment that determines which of the plants and animals survive to be studied. Remember that the change in a plant or an animal could lead to a change in the environment that could cause a loss of that same plant or animal.

Even if we determine that there is a "direction" that plants and animals evolve towards, we still don't know where the direction will lead. Nor can we determine whether or not that place requires a change in the biological or physical part of the environment.

When we talk about directions, we are not even agreeing about what is changing. Biologists talk exclusively about living systems. But physical systems are constantly changing as well: daily, monthly and annually. There are even longer cycles for which we lack sufficient historic records to be able to recognize or speak to accurately. Then there are those who are concerned with cultural changes, political changes, social changes, personal changes, economic changes, historical changes, and even spiritual changes.

Debates Within Debates

Finally, let me discuss what biologists think when they talk about whether evolution can have a direction or not. Generally, biologists do not think that an individual organism evolves. The individual organism goes through its life cycle and may change in its appearance as it goes through some predictable sequence of stages. But that is not considered evolution.

Instead, biologists believe evolution occurs within populations of organisms. The offspring of one generation may appear slightly different from the previous generation because certain characteristics were favored by the environment. So the question for biologists becomes "Is there a continual improvement in the population that leads to more and more successful traits within every individual within the population?"

It is just as dangerous to speak for all biologists as it is to attempt to speak for all theologians. However, I think the great tradition has been to believe that evolution does have a direction, although mutations introduce random components.

Theodosius Dobzhansky (Dobzhansky, 2013), an influential evolutionary biologist, wrote a paper in 1954 entitled "Evolution as a Creative Process". He randomly placed fruit flies into vials and let them breed as they would. Through multiple generations, the populations became more productive. There is a lot of research on this concept based on mathematical models and computer simulations, as well as actual breeding experiments such as Dobzhansky's.

Joan Roughgarden published papers in the 1970's showing that evolution generally resulted in larger, presumably better-adapted, populations (Roughgarden, 2006). She also showed that higher populations of one species could sometimes harm other species that were in competition for the same essential components. Conversely, there were some species that helped other species to increase their numbers.

This happens because the environment, the factor that does the selecting, is also constantly evolving. It is changed by the success of any given species. If evolution always led to continued progress, there would never be extinctions since animals would always evolve to be able to thrive in their environments. But because changes can also be detrimental, evolution hastens the demise of some populations when the environment changes rapidly and radically. Evidence from the fossil record demonstrates this process.

Not all environmental changes are manmade. Natural changes, or changes induced by the success of certain offspring, lead to the eradication of other organisms or traits. This, of course, affects the entire concept of a species since there is no population that remains unchanged.

There are biologists who believe that evolution has led to many different successful organisms which have come and gone with time. Compare this to looking down upon a mountain range with its peaks and valleys. Some animals increase in abundance to a peak, while others decline in a valley. There may be many other peaks, relatively taller and shorter, that represent numbers of individuals in various successful populations. Less-successful species are represented by the valleys. Each time period is a snapshot. Another picture, taken later, would probably show even different peaks and valleys.

So what? Actually, so nothing. You have simply become privy to a bunch of academics' arguments. And academics have evolved to argue, so that is nothing new! The point of this discussion is to show that what biologists talk about, when they talk about change, is not necessarily the same thing that Christians talk about when they talk about change.

Some biologists, like Dobzhansky, almost see evolutionary progress from a spiritual perspective. Others see the direction of evolution as a possible fact, but they see it leading nowhere. Still others see no direction to evolution. So when you hear a biologist, such as Richard Dawkins, make flat-out statements about evolution and its direction in their arguments against religion, you can be sure that they are ignorant, ill-informed, or intentionally misleading.

Evolution and Religion

I hope it is obvious, by now, that evolution and religion are not automatically found at the opposite ends of a spectrum. There is much opportunity for scientists to entertain religious thought concerning evolution and for religious people to envision evolution as a part of Gods purpose and design.

Whether or not God is involved with evolution is neither supported nor opposed by science. Many of my colleagues have strong opinions one way or the other. Many more have no opinion about these matters and tend to remain silent when opposing views are expressed. Just so you'll know, I see no problem with envisioning evolution as a part of God's design for his creations in the fullness of His time. I don't pretend to know what God's plan is.

Chapter 15 - Individuals

I don't know . . .

I don't know what goes on in a beehive at night even though I have been keeping bees for a long time. Bees don't sleep. One of the interesting things about honeybees is that the bees don't exist out of the hive. Of course, they leave the hive to forage, although only a few of them even do that. But if they are taken away from the hive, they don't survive long regardless of whether or not they are provided shelter, food, and water.

This raises an interesting question. Is the individual bee the organism, or is the hive the organism that simply has parts which can become temporarily disjointed? This question is not my original concept. There are numerous biologists who have proposed that the hive is an organism, and the bees are merely collections of cells.

This thought has interesting ramifications for our world. Bees are the predominant pollinators on the earth. Pollinated plants are the predominant life form and our major source of food. Without pollinators, there are no cheeseburgers! Yet, recently, one third of the honeybee hives have been lost each year. The loss is probably due to many factors, most of which are unidentified. Can you imagine humanity's reaction if one third of the cattle had died? And cattle don't pollinate anything.

Through human medicine, we know that while disease may have an effect on an individual cell, we usually make diagnoses by examining the whole. We listen to the heart, take blood pressure, analyze bodily fluids such as blood or urine, and even measure, calculate, and analyze our breathing.

Today's scientists seem to concentrate on the individual bee, and its physiology and behavior, to determine why the bees are dying. Is it the bees or the hive that dies? We know very little about what healthy hives sound like, what kinds of fluids they might have within, or the volume and quality of the air escaping from the hive. Until we decide whether it is the bee or the hive that is sick, we cannot know the cause of their demise.

We have somewhat the same problem when studying evolution.

What is an Individual?

This is actually two questions. What is individualism? And what is individuality?

"Individualism" is the idea that everything that happens to an organism is due to individual effort and behavior. There is a strong social current in the United States for individualism as it applies to people. Successful people like to believe that their successes and survival are creditable to their talents and efforts. There are many successful people who give talks and write books telling others how hard they worked and how smart they are. They say that others can do it also if they are willing to listen to their talks or read their books.

Of course, it is impossible to know how many people may have been "as smart" and worked "as hard" but failed because of other circumstances. No one wants to listen to them or read their books because they did not succeed. Cemeteries are full of these men and women. Of course, they are full of the successful ones, too, since the same reward awaits us all regardless of how we define success.

Well, this concept applies to plants and animals in a way that scientists don't have a way of taking into account. Is an organism's breeding and reproductive success dependent entirely on its own efforts and qualities? We generally assume that they are.

Yet in the case of bees and flowers, this can hardly be true. Does the brightest and biggest flower get the bee? Of course not. Brilliant flowers can bloom where there are no bees. Does only the fastest, most-aggressive bee get to the flowers? Well, no. Fast and aggressive bees may find themselves in a place where there are few flowers. In either case, reproductive success does not depend entirely on individual characteristics alone.

Neither does reproduction in mammals depend on individual characteristics. Predation may decrease the number of males so that, irrespective of how attractive the female might be, she may not find a mate. Or there may be too few females available for each male. This would lead to the same reduction in individual abilities to find a mate.

In other words, as mentioned before, the environment has an influence that is generally ignored by evolutionary biologists. But it can greatly alter the chances of an animal reproducing successfully. When reproductive success depends on the behavior of others, biologists have no way of apportioning the credit.

"Individuality" is a problem we have with bees as well as many other organisms. What exactly is an individual? All bees come from a single individual in the hive, the queen. Their appearances and behaviors are altered by a combination of genetics, environment, and experiences.   But they are all descendants of the same individual. This may seem like a fine distinction, since we are used to conceiving of individual units as distinct individuals. However, the concept of "individual" gets murkier.

For example, "What is a grape plant?" Grapes are commonly reproduced by sticking a stem of a plant into the ground until it takes root. So if we put one hundred stems from the same plant into the ground to create a vineyard, is the individual a single plant or is the vineyard the individual plant?   Are they all grapes from a single individual now in many different places, or is each plant an individual? What is a grafted plant such as an apple, peach, or rose where the vegetative stem is from a different individual than the root stock?

There are thousands of plants and animals that reproduce by simply dividing themselves in two. This raises a problem for evolution because we don't know if we should treat an amoeba as if each cell produced is an individual or whether all descendants of the cell are the same cell. Then there are those animals, such as the Portuguese Man-O-War, that only exist as colonies. How does one assess the reproductive success of the individual cell within the colony?

Traditional Darwinism emphasizes individuals and populations. It does not take into account, for example, that wolves are social animals, and a given wolf has a greater chance of reproduction if it is part of a successful pack. In other words, the individual's success is not entirely dependent upon its own merits. Breeding success is not always an individual characteristic.

In fact, the amount of attention devoted to evolution, the supposed central theme of all biology, has distracted biologists from the study of other concepts that may be equally, or more, powerful ideas in the biological world. Evolution has emphasized the success of the individual which in turn emphasizes the role of competition between individuals. This is what gave rise to the still influential role of social Darwinism and the idea that all behavior is acceptable for the survival of the fit.

The survival of the fittest has been commonly portrayed as "the fastest and meanest, bloody tooth and claw". Yet biologists have not come to grips with the even-more predominant roles among living things of cooperation and inter-relationships.

The Most Common Mode of Existence

Every living thing, to be carefully examined, has been found to have at least one other organism living on or within it. Most have at least one organism that is found exclusively on the species as well as a few others that can be found on other species.

Consider this. If every living thing has at least one other species living on it, then living together in some kind of biological relationship is the most common form of existence. The obvious and inescapable truth is that living things live on living things, and such relationships are the most common form of life.

We have coined various names for these associations such as commensal, mutualist, or parasitic. But these categories of being are almost always couched in terms of which animals are harmed or benefitted. Yet "benefit" is not a scientifically-verifiable quality.

Most of these associations are comprised of very confusing relationships in that they are sometimes harmful and other times beneficial. After all, harm and benefit are simply synonyms for good and bad which, within themselves, are value judgments. Value judgments are not material and therefore cannot be used as an efficient means of scientific definition.

A better collective term for these organisms is symbiont. This is an ecological term which simply means "living together", or "one that lives in association with another". Such a term even includes predators and herbivores. After all, one does not find deer without grass, or lions without zebras.

Evolutionary biologists have extremely poor ways of talking about the overwhelming success and abundance of these associations. In the case of so-called "parasites", they are almost always defined as free-living animals that have lost their abilities to live independently, or have degenerated to a parasitic state. Are we to believe that half of all living things have evolved backwards to be associated with only certain other organisms?

A Thought Experiment

Instead, using sequence and logic, imagine the first cell. A cell is conceived of as an individual unit that can make a copy of itself. Most biologists agree that the first cell probably developed in water. I believe it says that in Genesis also. Water seeks the lowest level, so the cell most likely lived in a pool of some kind, albeit perhaps a large pool like the ocean. Further, most cells today do not live floating free but live on a surface. So we might assume that this primitive cell lived on the bottom of the pool. Note that "initial assumptions always make a difference".

The point is, if that cell repeatedly divided in two, it would eventually have covered the bottom of the pool. Then, if the cells were divided into two again, there would be nowhere for the new cell to go but on top of another cell. Two cells living above and below each other are going to have an influence on one another. The result could as likely be a cooperative, multicellular organism, as it could be two organisms living in close association: a symbiotic relationship.

Yet the role of parasitism in evolution, using even its parasitic definition, has not been adequately addressed. The entire cooperative and interdependent nature of living systems has never been adequately explained, or even recognized, by many evolutionary biologists.

So are these two associated organisms, or cells, an individual or a unit? Darwin talks about the survival of the fittest. Evolutionary biology, however, does not account for the fact that the word "fit" may not mean the biggest and meanest but the most cooperative and symbiotic.

In fact, since all of life shares the same genetic material, perhaps life itself is the whole organism, and there are no individuals. That is an extreme interpretation, of course. I bring it up to demonstrate, though, a major problem in evolutionary thought that does exist. Of course, biologists might one-day address this issue of symbiotic relationships, but so far they have not.

The question of what constitutes an individual, whether viewed as individualism or individuality, remains undefined. It would seem to me that the logical, defensible, scientific definition of an individual would be the recombination of two sets of genes. This process could occur as in the case of sexually producing organisms or as in the reconstitution of a new set of genes in organisms that reproduce asexually. Both proposals have cultural and political ramifications concerning abortion that are rejected by many scientists for various reasons.

At the same time, there are those scientists who present evolution and science as completely proven facts that disprove all religion. Admittedly, there are many unanswered questions. Religion might provide needed insight into answering them.

Unanswered Questions

That there are unanswered questions does not disprove evolution. However, these questions are purposely ignored by Richard Dawkins and others who promote evolution as an argument against religion. Either they are woefully uninformed, or they are willfully ignoring and suppressing information about evolution in their discussions. It is hard to imagine any other explanation for their one-sided approach. If they are ignorant, they should not be consulted. If they are misleading, they should not be trusted.

The words "good" and "evil" can both be used to describe humanity. Science, the internet, art, music and mathematics can all be used in either cause. When science is wrong and it is used to mislead, it is no longer a higher mental activity or of benefit to society.

Chapter 16 - Problems Involving Religion and Science

I don't know . . .

    I don't know the answers to most questions that are at the center of the debate between science and religion. I do know that, contrary to the way many people want to frame the arguments, the answers are seldom simple, dichotomous choices. In every case, there are multiple possibilities and explanations that fit the data to one degree or another.

    It is the nature of humans to believe that their thought processes are correct. After all, these thoughts are usually based on all the experiences they have had up to that point including: emotional, unconscious processes and logic. Children of a certain age are convinced that tall water containers hold more water than short containers. They are absolutely certain of this fact until they are able to experiment and discover for themselves that it is possible for a short glass to hold more than a tall one. Then they change their world view, and their new discovery becomes just as vehemently believed in as their old one.

    The tendency of humans to believe they are right and have a perfect understanding of any situation is nearly universal. The Greeks call it hubris. It is easy to see why we so commonly fall into "hubris". After all, we cannot be aware of what we are not aware of . . . But we can consciously try to think of alternative explanations for circumstances and problems. We may reject many of our alternate proposals. However, in considering them, we may find two or three that we would have to admit are possible.

Instead, humans often decide that what they know and experience is what is true, and they reject any other explanation outright. This leads to the creation of opposite explanations as the only possible explanations when, in fact, there may be several of more-or-less equal validity. Sometimes new choices are superior to either of the "straw men" erected for us to fight over.

What are the essential questions that tend to divide religion and science? In the first part of this book, I gave alternative explanations for numerous religious and scientific views and theories. In this chapter, I hope to briefly discuss these central topics, of which there are four, that seem to separate us.

The Questions

The Creator Question: Is there a creator of the universe, or does the universe exist by chance interaction of existing circumstances? In spite of the fact that theologians and scientists have numerous explanations for how the universe may have come into existence, they cannot prove any of them.

It is not unreasonable to believe scientific explanations, but each explanation has problems or inconsistencies. In fact, almost all scientific theories are based on certain initial assumptions which may or may not be true. Theologians get around the same problems by invoking faith as a means for not needing to prove their theories. They basically throw up their hands and say, "We can't prove this, but we believe it."

The scientific process provides a means for observing and asking questions of the material world. The method is based on empirical and measureable evidence. This restricts scientific inquiry to the observation of the material world which is definable and measureable. Religion, by contrast, deals with non-material subject matter which cannot be quantified. Therefore, a religious experiment cannot be duplicated, measured, or compared.

This does not mean that there cannot be spiritual truth, as well as ways of knowing, that are extremely important. Freedom, love, patriotism, ideas, and numerous other abstract concepts are extremely important though not material or directly measureable.

At the root of both sides' arguments, are assumptions that must be accepted on faith! The existence of a creator cannot be proven or disproven. In fact, the entire concept of proof and disproof is in question for two reasons.

First, humans do not agree on what a creator is, and/or how a creator would function. It is difficult to prove any concept which cannot be defined. The scientific definition of a creator, though, is usually not the same as the religious definition.

For example, scientists accept on faith that the laws of the universe have always existed and are uniform throughout time and space. However, they have no explanation for why those laws exist in the first place. What is the source of these infinite laws? Could the source be called God? Similarly, scientists assume that some kind of matter has always existed for the laws to act upon. But has one actually explained the origins of anything if the source of original matter is not accounted for?

Secondly, the entire concept of proof is undefined. For any proven fact, only one exception needs to be found before we realize that, what we thought was proven, was not at all. Scientists deal with this all of the time by revising their proofs. In fact, they take great pride in the fact that, as we understand systems better, we revise and update our theories. However, this leaves them never knowing for sure if their most recent conclusions are truly proven or not.

In short, neither religion nor science can really prove there is a God, or that there isn't. Those who claim to have done so are either ignorant, misled, or purposely misrepresenting the case.

I hasten to note that not all science is Godless. There are numerous, believing scientists. They tend to be somewhat quiet for a variety of reasons. For one, they recognize the inherent weaknesses in Godless claims, so they see no sense in making God an issue. Further, because of the hubris and intolerance of other scientists, they recognize that delving into such discussions would only invite vitriol, anger, and personal attacks.

Very few religious people reject all science. They simply see science as evidence of God's hand in the world. Sometimes they even seem a little bemused by scientific excitement when something is discovered scientifically that they already believed through faith. However, most religions do not impose the burden of proof on themselves and are in a weak defensive position when confronted by so-called scientific proofs.

Lacking in the discussion of a creator, generally by both parties, is a simple acknowledgement of the fact that both beliefs may be wrong, or that part of each answer may be right.

The Creation Question: The creation question often blends in with the creator question, but it is distinctly different. The creator question concerns initial causes; that is answers to what caused the events leading up to our current state of the universe. The creation question concerns methodology, or questions about how the universe was literally put together.

Religion has little, or no, explanation of methodology. Instead, for the most part, religious believers rely on science to explain the methods. Science is full of methodological explanations about how the universe was created, none of which can be verified.

Because religion cannot explain how things happened, religious believers often succumb to vague, magical explanations. For example, when it is said that God created the earth in Genesis, they do not have mechanical explanations for that. But they certainly do have a sequence of events that acts as an organizing order. It is interesting that most religions don't even address how the universe is created. They simply believe that God did it. I would guess the average religious person more or less thinks that God waved his hand and spoke the magic words. Perhaps that is exactly what happened.

I discussed the sequence proposed by the only religions to actually put forth any detail: Judaism and Christianity. Their explanation of sequence closely follows the thinking of how events might have occurred as proposed by the scientists. The difference is that religions explanation was first recorded thousands of years before any scientific theories for the event existed.

Scientists tend to think that laws and matter have always existed and reacted together in ways which are universal in time and space. Yet, unexplained is how these laws and matter were set in motion in a series of hypothetical reactions resulting in today's world. While the sequence and mechanical steps are elaborately explained, they remain unverifiable and contain many inconsistencies.

The Life Question: When does life begin? When does life end? For most of human history, these questions have been non-issues. Life began at birth and ended at death. Even in the ancient past, however, it was clear that something was alive within the expectant mother. Motion and heartbeat were indications that the baby was alive.

With further scientific understanding of reproduction and the fusion of gametes to produce offspring, a debate has developed about whether or not a fertilized egg is alive. The answer to this question is clouded by philosophies, human desires, politics, economics, opinions, and greed. It might be more interesting to consider how humans decide someone is dead.

177

Breath. Early humans probably first noted that dead animals and humans do not breath. The words *pneuma* and *spiritus* both refer to the breath of life, that magical thing that ceases upon death. Modern science has been able to show that temporary cessations of breathing need not be fatal, and those periods of time may be very long depending on temperature and other conditions.

In the case of human embryos, we know that they do not breathe. However, we also know that breathing is essential for the embryo's well-being. They simply rely on their Mothers to perform this function. One of the legal ways of determining whether an embryo is still born or not is to check to see if the lungs were ever inflated. If a small sample of lung tissue floats in water, we can be assured the lungs were inflated; and the child was born alive.

Since an embryo's lungs are not inflated prior to birth, this can be an argument that they are not alive in the womb. This seems a slightly manufactured and thin definition given the other physiological facts.

Blood. It has been known since written history and before that dead animals do not bleed. The blood coagulates in the body when the heart stops and no longer flows rapidly from a wound. Every butcher knows the need to bleed out freshly-killed animals as quickly as possible so the coagulated blood does not make the meat unpalatable.

Human embryos have flowing blood at very early stages. The heart is beating by six weeks, and presumably something is being moved. Using only this criteria, the human embryo could be said to be alive at six weeks.

Heart. When the heart stops, the blood ceases to flow and coagulates. Death quickly ensues. The cessation of the heartbeat has been a measure of life since time unrecorded. In using this same criteria, we could say that life begins at six weeks. As the embryo grows larger, it cannot meet the nutritional needs of all its cells without a circulatory system.

Six weeks comes very early in a process that is forty weeks long. It is also the period of time when most spontaneous, natural abortions take place. In many of those instances, women may not even be totally aware that they are pregnant.

Brain waves. More recently, humans have defined death in relation to electrical signals from the brain. When these become abnormal for long periods of time, or are absent, we say that the person is dead. However, experience has also shown that respiration, blood flow, and heartbeat can continue indefinitely when the brain is more or less dead.

While there is a lot of information supporting the idea that brainwaves begin in the embryo as early as six weeks, reliability seems questionable. The more reliable estimate is that brain activity begins sometime after twenty weeks. Even then, these are not typical, adult, brainwave patterns. They show developing activity that will change dramatically and increase in complexity over time for, literally, many years.

Those who have observed death in any animal cannot help but be impressed that something undefined happens at that moment. It is more than the cessation of heart beat or respiration. Something difficult to describe either leaves the animal or overcomes the animal.

This change has not been identified or defined by science. It seems to be a subjective and abstract event. Science cannot investigate what cannot be defined and, therefore, cannot contribute more to the discussion of death than regarding outward symptoms.

If we cannot isolate and define the changes that occur at death, then we certainly cannot add anything to the discussion on how life enters the body. These are not scientific subjects no matter how much testimony one side claims or how much science another side may use to back up their arguments.

The Gender Question: Gender may seem like an odd topic for this book. However, religion almost always has something to say about gender roles. Science is also continuously used, correctly and incorrectly, to justify positions concerning gender identity and roles. The war between the sexes has not always existed. The modern perception is that it has, and often historical anecdotes are used to justify positions on either side.

Through much of history, though, the arrangement between men and women was mostly cooperative. Of course, that does not imply equal. Men most often dominated major decisions, women were seen as possessions, and many opportunities were withheld from women. However, women were also given much free rein in running the children and household and had many opportunities that the men did not have. My grandmother had complete control of the chickens and eggs though she lived in a male-dominated household. After my grandparents' deaths, it became obvious from examining their account books that she had far more income than my grandfather.

While there have always been experiences of domination and dishonesty from both genders, the most common arrangement between the sexes seems to have been for a man and a woman to marry. The woman then cared for the home and family. The man labored daily and turned the greatest portion of his income over to the woman for those purposes. This has been an amazing act of voluntary income redistribution that seems to become the accepted norm!

The war between the sexes actually began in the 1800's and, once again, under the influence of Charles Darwin. One of the last parts of his theory of evolution is called "sexual selection", and it is a continuation of his theory of survival of the fittest. Unfortunately, this part of his theory is the least scientific and had very little actual research behind it.

He was concerned with how mates are chosen and such oddities as the peacock tail. He proposed the theory that male ornamentation existed to impress the female. Unfortunately, his theory about these things has come to be viewed as an all-encompassing theory of gender roles.

The problem with his theory is that it has created a thousand generalizations, two thousand exceptions, and numerous special cases in the real world that have to be explained away. It isn't even a theory. It is an idea that every exception disproves.

Here are some quotes from Darwin's book, *The Descent of Man*, published in 1871 (Darwin, 2004).

> "Males of almost all animals have stronger passions than females . . . the female, with the rarest exceptions, is less eager than the male."

> "Females want mates who are vigorous and well-armed."

> "Many female progenitors of the peacock . . . must have . . . by the continued of the most beautiful males, rendered the peacock the most splendid of living birds."

This gender concept is so completely ingrained on our modern culture that it sounds like it must surely be true. If this is what one is taught in biology class, then it seems unreasonable to question it as scientific fact. Of course, if one reads the statements critically, they certainly don't sound scientific.

Beautiful peacocks are not necessarily good fighters, and there is no evidence that females of any animals really care about with whom they mate. Often the winner of a male contest gets the whole herd. The individual females don't get to choose. In addition, there is no evidence that we share the female peacocks' taste in beauty. The birds may have been colorblind for all Darwin knew.

But the real problem with Darwin's theory of "sexual selection" is that it assumes males and females have innate, basic, and immutable conflicts of interest. The war between the sexes had begun! According to Darwin, females want mates with good genes. Males want a lot of sex. There is a necessary and inevitable conflict in the theory.

I know Darwin's concept of sexual selection revolving around conflict of interest is accepted dogma in modern culture. The serious problem is that it isn't true.

Biologically, males and females begin with the exact same goal. They want to have offspring. From the very beginning, males and females are involved in a joint venture. It is not self-evident that the female has less passion than males. In fact, there is little evidence to support any of Darwin's ideas about gender.

If conflict is the initial relationship, then negotiation and reconciliation must continuously follow. However, when we examine actual animal behavior, we see far more cooperation than conflict at the beginning. Conflict may develop, but the initial relationship is based on cooperation.

Many animals are capable of reproducing asexually. From a scientific perspective, sexual reproduction is an option, not a requirement. So why does sex exist at all? The biological purpose of sex is to share genes. The combination of genes creates new individuals with a high likelihood for success, without the risks of mutation, but with a new mixture of traits, that may prove beneficial to the species. It is a cooperative venture.

The biggest problem with Darwin's theory of gender roles is that it simply doesn't describe what actually exists in the animal world. In many species, the sexual behavior roles are reversed, with males nurturing offspring. In other species, the two genders look exactly alike. In still other species, the females are the more colorful. There is even little evidence that females are less passionate. In many species, that does not appear to be the case. What about the species that undergoes sex changes as a normal course in their life cycle, changing from male to female or vice versa?

Who cares? Most religions emphasize the sacredness of the family. Most of our common mores concern the proper care of the young. Modern secular ideas have usually been supported by the Darwin theory of "sexual selection". They use competition between the sexes as the beginning point of discussion. Religion, on the other hand, usually views the male/female relationship as a shared experience devoted to the perpetuation of the race through the bearing of children and the perpetuation of the culture.

Of course, if Darwin is wrong, todays "war between the sexes", feminist agendas, and female liberation make considerably less sense.

Chapter 17 - Literal Bible

I don't know . . .

 I don't know why some people make a big deal out of Bible literalism. Sure, there are those who accept the Bible in a literal sense and deny reason and reality. But that is not a blanket condemnation of the Bible or of religion. There are scientists who have used science for wicked ends and who have lied and cheated as well. That does not make science wrong. Perhaps we should examine what the Bible actually says about science and living things.

 Contrary to popular belief, I am not old enough to personally remember Adam and Eve. I've never even met Abraham. Theirs are very old stories. I am told that Hebrew provided one of the very first written alphabets. I am also told that writing and numbers came into existence almost simultaneously in many cultures around six thousand years ago. I'd like to meet the guy who remembers that. But someone must know, or people wouldn't be saying it.

 Presumably, the events told about in Genesis took place before they were actually written down, which means they are very old stories indeed. Many people view these stories as myths. Then there are those who take the Bible very literally. People who believe the earth is flat usually base their ideas on Bible literalism, specifically according to the multiple references to the "ends of the earth." However, ancient sailors, which many of the early Christians were, have known the world was round for thousands of years. They based their understanding on the simple fact that when ships sailed away from land, the masts were the last part of the vessel to disappear from sight.

Flat-earth ideas are the kind of thing that make many people uncomfortable with Bible literalism. Many people see the Bible as a general guide to living a good life, but they reject any attempt to compare the Bible with scientific thought. I think I have shown that this needs not be the case.

On the other hand, if you are skeptical of one portion of the Bible, but not another portion, where do you draw the line? If the account of the creation is not literal in some sense, are the accounts of Matthew, Mark, Luke, and John literal or mythical? These are legitimate questions for believers.

The Bible has been translated repeatedly and copied by hand for centuries by many different peoples. Originally it was written in Hebrew, then translated into Greek and Latin before finally being written into The King James Version in English. This latter translation was done by a committee of men. Can anything good come from a committee?

Translation is a difficult process, and meanings can be lost and changed. I don't know how much of the Bible is literal and how much is metaphor. I have, however, arrived at a conclusion that I think is useful. I believe it is literal, as long as it accords with reality or the facts, as we understand them. That may sound like a self-serving approach, but I believe that is the approach that Jesus Christ taught.

Literalism

"Hear and understand. Not that which goeth into the mouth defileth a man; but that which cometh

186

out of the mouth, this defileth a man" (Matthew 15:10-11). In the very first teachings of the New Testament, Jesus taught that the rules, the literal dietary laws that had been in existence for centuries, were not what ought to be important. Instead, he advocated a common-sense approach based upon reality. It sounds almost rational.

"Let he who is without sin cast the first stone." "Render unto Caesar . . ." These are other examples of the way Christ adapted to reality rather than the literal. The clergy of his day were angry with him, but he repeated his analogies. He was very clear. It was not the literal rules of law, but the reality, that was paramount.

Jesus Christ did not arrive on the scene to say, "Here is a new set of rules." He didn't say, "I am changing the rules. It's now alright to eat pork." Instead, Jesus taught us a way of thinking and analyzing. In fact, his most common teaching method was with a parable, a story of how the world worked. Then he asked people to apply their understandings to their lives. You might say that his teachings were based upon reason, the foundation of science.

Would he be pleased to see people deny "demonstrated" reality in his name? I think not. Rather, I think he would adapt the demonstrated revolutions of the earth around the sun, unknown in his day, to teach a parable. Jesus wanted people to think for themselves. At the time, there were endless rules, laws, and punishments prescribed for the people.

Many of those ancient Judaic rules made perfect sense and continue to do so. It is probably not a good idea to sleep with your brother's wife or your uncle's wife. It is probably wise to watch what you eat and how it is prepared. These kinds of rules make up a great deal of the book of Leviticus. But as Christ taught, it isn't so much about "what one eats". It is more about our words and actions.

Jesus didn't teach with rules or laws. He taught through parables. Often he used the natural world and reality to teach his lessons. In short, he reinterpreted the scriptures of the Old Testament in light of current knowledge. "Think not that I am come to destroy the law, or the prophets: I am not come to destroy, but to fulfill" (Mathew 5:17).

One of the often cited conflicts about religion is the age of the earth. Geologists tell us the earth is very old. Biologists are not the ones who make that claim. However, for the most part, biologists accept the geologists' findings. They do this partly because of what they know about living things and partly because the past fits nicely with this idea. In fact, it was the blending of biology with geology that suggested that the continents might be moving apart. Today continental drift is an accepted idea in science with reliable documentation.

The number one concept of biology is that all of life is made of the same material and, therefore, can be said to be related. Secondly, living things change over time, even from one generation to the next. Neither of these ideas seems to contradict a literal interpretation of the Bible.

Modern geology claims that the earth was not created in six, twenty-four hour days. The scientific facts are clear, and the actual interpretation of Bible language and lack of detail leads me to side with the evidence rather than the literal. I think that is what Jesus usually did.

I am aware that there are people on both sides of this great debate who believe that this "theistic evolution" is an unprincipled compromise, an attempt to have it both ways. They think such a conclusion represents "moral sloppiness". On the other hand, true scientists would call their logic "sloppiness".

Both positions strike me as unreasonable and illogical. First, to literalists I would say, "If God created the world, and continues to do so through natural processes, why would he suddenly jump outside those processes that He, Himself, created to create life?" I see no evidence in the Bible that shows God would prefer to work outside of natural processes. In fact, if that were required, it would imply that the natural processes that He created were inadequate for His own purposes. By definition, He can work outside such natural processes if he so chooses. However, if God saw his creation as "good", as it appears He did in Genesis 1:12, how could those processes also be inadequate for His purposes?

To the scientist, I would ask, "Why is it necessary to circle the wagons and fight to the death over concepts that cannot be verified or falsified using scientific methods?" Scientists are quick to point out that they deal with the natural world and reality. Things that fall outside that realm are not

189

anything science can shed light on except as in personal opinion.

How can we be sure of scientific conclusions when there are so many unanswered questions remaining? What is an individual? Animals and plants work, not only as individuals, but also as members of a group within an environment. How does the changing individual change the group and the environment? And what effect does the environment then have on the changing individual?

Human hubris has been shown to be wrong as frequently as religion. By definition, God can do as He pleases. He can reveal Himself, or not, as He pleases. Whether it seems logical to scientists or not is completely irrelevant. If there is a God, something none can state of a surety, He is not required to make sense to scientists, or theologians, in our narrow understanding of this place and time.

Might it be that neither the theologians nor the scientists are correct? Of course. Or perhaps they are both right, but our information is yet too incomplete for us to understand. The war, as perhaps with most wars, is sort of ridiculous.

In addition, it must be recognized that not all science is godless, nor is all religion anti-science. There are many imminent scientists who have varying degrees of belief in deity from the agnostic to the devout. There are many theologians who are intelligent and well-educated who accept the reliable findings of science as revealing more about the nature of the creator.

Part 4 - War

Chapter 18 - Reasoning About Reason

I don't know . . .

I don't know how I think. I know I do think, but the process is quite obscure. Humans have described and classified certain ways of thinking. Classification is a type of thinking in itself, I suppose. We have come up with all different kinds of words for how we think we know things: authoritarianism, rationalism, empiricism, pragmatism, mysticism, and so on.

It is important to understand at least a few of these classifications of thinking because science, especially, relies on these as being foundational to the process. That isn't to say that religion does not also utilize many of these methods. However, religion also relies on other methods of thinking and knowing which science rejects. Could there actually be a scientific basis for some of these other ways of knowing?

Philosophy

Rationalism is a process of thinking where humans certify, or accept, an idea because it either agrees with, or is deduced from, things we already believe. The theorems of geometry are an example. The weakness of this approach is that we must begin with a set of beliefs. This set of beliefs may not be rational themselves. Scientists like to believe that "theirs" is the rational view. They have their own set of fundamental and unexplained beliefs.

Science often relies upon <u>empiricism</u>, the use of the physical senses, to verify ideas. There are several shortcomings in this approach that science usually ignores. First, this approach relies on a stockpile of experiences with which to compare results. When you ask someone to see if the light switch is working, you assume they know what a light switch is. Perhaps, even more to the point, our senses cannot sense all important things. We cannot sense the past, yet science is based upon past findings. We cannot sense very distant things, very tiny things, or even many things that other animals are able to sense. For example, humans do not possess the senses for echolocation that bats seem to have.

<u>Skepticism</u> is the rejection of all ideas that do not have sufficient physical evidence to support them. It is actually the basis for all thinking whether religious, scientific, or philosophical. Of course, it is possible to be too skeptical thus rejecting all patterns of thought. However, if one is not skeptical enough, he can be easily duped.

People are sometimes wary of <u>authoritarianism</u> which is the acceptance of ideas based upon communication with another person. Science often portrays religion in this light. The greatest weakness of this method is that people must determine which authority to trust. Though sometimes authoritarianism is portrayed as a bad thing, I would guess that a huge amount of what we each think we know, "we", including scientists and theologians, accept on authority. I have never personally been to Africa, but I believe it is there.

Pragmatism is the acceptance of an idea as true because it works for you. People created strong and elaborate steel and iron objects long before the principles of physics and chemistry were understood. Rational clarification is not always necessary to do many practical and important things.

In fact, in some areas of activity, there may simply be no scientific or theological basis to back up belief in an idea. The weakness of this method, of course, is that something may only seem to work while, in fact, there was an erroneous correlation with some other event. We may also find that things that work in the short run have long-term, poor consequences. Science, itself, repeatedly falls into this trap. An easy case in point is in the use of medications which may prove beneficial in the short term but cause long term problems.

Mysticism is often associated with religion. Generally, mysticism rejects rational thought for feeling. It appears to abandon the mind. But for a true mystic, this may not be a problem. They do not require a consistent set of ideas to find peace in the world. We will see that some scientists also behave in this fashion as well.

Everyone thinks they "think", and most of us think we "think" pretty well. Scientists, especially, seem to try cornering the market on thinking by claiming that no one else thinks, or at least thinks as clearly as they. Unfortunately, theologians have ceded this claim to a great extent by talking about feelings and spiritual ways of knowing that don't seem to have a basis in thinking.

Perhaps, with so many competing ideas, we should examine what actually happens in the brain when we think. Where does thinking happen? What is actually happening when we think? Scientists have claimed that we are nothing more than an electrical and chemical machine, and that is all that thinking entails. But they have not actually had much to say about how we go about doing it. Perhaps thinking isn't what scientists think it is.

The Brain

We haven't known much about how we "think" until late in the twentieth century. Since about the 1980's or so, there has been a revolution in brain science that has allowed us to better understand the structure and functioning of the brain. Primarily this is because of new imaging technology that has allowed us to examine what goes on inside the brain. To some extent, we understand the brain in ways that were not possible before.

Theologians have not had much to say about the advances in brain science. Their concerns are often more philosophical dealing with issues of behavior such as ethics, morality, pain, suffering, and relationships with others, including deity.

It is instructive to note that, while science itself has contributed to our understanding of the brain, some scientists have argued this as proof that we are actually just automata, obeying physical laws of chemistry and physics. Conventional science has long held the position that 'the mind' is merely an illusion. Material science and neuroscientists have assumed that the mind is the result of collective, electrochemical activity of the neurons. However, newer work has reached the opposite conclusion. It is that the human mind is an independent entity that can shape and control the functioning of the physical brain.

Both conclusions raise questions about scientific claims of reason and logic, since reason and logic are the result of electrochemical activity in the brain. Can I trust reason and logic knowing that it is just the product of chemical reactions?

Functional Units

The brain is made of special cells called neurons. They are special in that they have the ability to pass information along their membranes as very weak electrical signals. They also have the ability to pass these signals from cell to cell.

It is a little more complex than that. Not only do they pass the signal from one cell to another, one cell can pass the information to multiple cells at once so the signal can spread very rapidly in many different directions. This is sometimes called "parallel processing" meaning the information can be sent to numerous destinations at the same time or in parallel. Since each neuron can be connected to numerous other neurons, the collection of neurons resembles a complex, three-dimensional fish net with information traveling along many different pathways simultaneously. Thus, from the point of origin to the desired destination, there are many different pathways possible.

When a child is born, it has about twenty billion neurons in its cerebrum. The cerebrum is comprised of the top six layers of the brain where consciousness exists. Each of these neurons has about one thousand connections to other neurons. By the time the child is ten or twelve years old, each neuron has about 10,000 connections to other neurons. That is one complex network with an infinite number of possible pathways for information to follow.

This means, of course, that there are an infinite number of ways of thinking just at the conscious level. What are the implications of this? As incoming information is passed from its point of origin, say the fingers, it does not have a single prescribed way of traveling to its final point of consciousness in the brain. Even if the "correct" point is arrived at in the brain, it does not mean that the same pathway was utilized by each thinker.

This explains why humans tend to be less accurate than computers at precise thinking. Computers have only a single pathway to the correct answer. Thus, they are extremely accurate in precise calculations. If the wrong path is selected by the computer at any intersection, we get either a complete shut down or wildly incorrect answers. When humans make an error in calculation, we may not be correct, but we are usually close in approximation.

So the scientific explanation of how the brain functions at the simplest level calls into question the reliability of clear thinking of any kind. Because all this happens automatically, below our consciousness and control, we cannot be assured that conclusions reached by reason and scientific data, alone, are reliable. That is why, in the sciences, we ask for replication of data and some ability to predict and control material events as evidence of clear thinking.

Yet some scientists have drawn far-reaching conclusions without data that leads to the ability to predict and control events. Evolution is a key example. Without questioning the theory, I can tell you that the theory does not predict what will happen in nature. It only allows us to control the results in very limited, artificial, closed systems such as agriculture or research. This does not mean that science, as a whole, has failed these tests. Many scientific advancements have come about because of our ability to predict and control.

Of course, religion has the same limitations to clear thinking and the inability to accurately predict and control outcomes. However, many of the tenets of religious living can predict, in a rough manner, an arguably more fulfilling and peaceful life, especially in relationship to promoting community.

Compartmentalization

The brain is a compartmentalized and layered structure. Understanding this structure is essential to understanding how thinking works. The structure of the brain has many ramifications on how we think.

The brain consists of at least four layers of neurons. Then there are specialized areas or compartments associated with the various bodily senses. First let's identify the layers.

Receptors: These cells, found in the periphery of the body, are not always considered a part of the brain but are continuous with our brains. It is in these cells, in our senses, where much thinking begins. When we feel, hear, see, smell, or taste something, a signal is sent to our brains.

While these signals are more or less mechanical and extremely similar between individuals, they are not necessarily identical. It is probable that some individuals have a slightly different number of specialized cells for color vision, for example. This difference could explain the arguments over shades and nuances of color that exist between people.

Probably no two persons' perception is identical. The best we can hope for is a shared acceptance of our experiences, and there is even some question about conclusions drawn from experience. These physical experiences are foundational to science and are another reason why scientists insist on duplicated and quantitative data that leads to prediction and control.

Variations in physical experiences are also claimed by religious people as foundational to spiritual manifestations. In this case, unique experiences are often attributed to certain people as manifestations of the spirit. These can neither be proven nor excluded by scientific or religious investigations.

Brainstem: The inner-most layer of the brain proper begins with the spinal cord and brainstem.

This is where the neurons coming into the brain are first sorted into pathways routed to and from different higher areas of the brain. We have no conscious control of these functions. Incoming signals, and the pathways they follow, are hardwired into each individual and happen without given permission. Again, there may be variability from individual to individual, just as there is variability in body shape, build, skin color etc.

This is where wires cross from one side of the body to the other side of the brain. In general, the right side of the brain controls the left side of the body and vice versa. This is also one place where information can be shunted into numerous directions simultaneously. Some signals can be sent to different portions of the brain, such as the pain center, at the same time it is being sent on to other areas such as the conscious layer of the brain.

Midbrain: The midbrain also functions below our conscious level. It contains numerous specialized compartments that perform functions which we have had difficulty identifying until recent years. We are still discovering some of these areas and often have only vague ideas about what they do.

One area of this region is an area called the thalamus. This is a center which somehow interprets pain and pleasure. It's an essential learning center in many ways. Children cannot know to avoid hot until they know what hot means. It is the thalamus that teaches them "hot".

Other areas in the midbrain are related to emotions and memory. Generally, we have no conscious control over the midbrain. In fact, we have no consciousness of the events that are taking place there. This means that much of what arrives at the area of conscious decision making and response has already been sorted and influenced in important ways.

We know the least about how the brain functions in this area. We know that the neurons here all utilize chemicals and electrical signals as in other parts of the brain. However, many of the chemical transmitters may be different, and even more signals can be generated to other areas of the brain from here. Not all human, mental, activity is based on reason. The midbrain presorts incoming signals in ways that we do not yet understand.

For example, the soccer player who sees several team mates running on the far side of the field, while he himself is running and dribbling a soccer ball, has but a split second in which to decide which team mate to pass to. These decisions are not based on reason. There is no time for reason.

Driving a car in traffic and waving to a friend while talking on a cell phone is not done by reason. Sensing that something is troubling your spouse or friend is not often a reasonable thought. In fact, most thought that makes up our daily activity and survival is not based on reason at all, but is predetermined by emotional and perspective decisions that are made below a conscious level.

There is a region of the brain called the orbitofrontal cortex. You might note the name. Orbit refers to the orbit of the eye, and frontal refers to the front part of the cerebrum. Apparently this area of the brain is responsible for integrating visceral emotions and responses into the reasoning process, frequently over-riding logic and reason.

Because these kinds of brain activity operate below a conscious level, neither religious believers, nor scientists, can know for sure if they are influenced by emotional and visceral reactions other than reason.

Cerebrum: The cerebrum, mentioned previously, is the only layer of the brain that is conscious. But it is not even clear what portion of this organ is actually under conscious control. It is made up of billions of interconnected cells that signal to each other. There are six layers of cells, each about the thickness of a playing card. Stack six cards together, and you will have some idea of the thickness of this thin layer of consciousness that rests atop the brain.

How much of the bottom layer we can alter with our thinking and experiences is not clear. We obviously can learn many things and restructure connections in this layer to a large extent. But we are not clear on "how", or "where", "thinking" or "learning" occurs.

Our reasoning ability is even more restricted than you might think from this brief overview. As you have seen, only the cerebrum is used for reasoning at all. This means that three fourths of the nervous system is under automatic control by all kinds of influences which are still mostly unrecognized or vague.

Even the cerebrum is not entirely for "thinking", at least as we think of thinking. The cerebrum consists of five different lobes. All but one is wired to a certain modality or perception. The back part of the brain is where we normally perceive sight. The part of the cerebrum that lays approximately above the ear is where we experience hearing and control speech. The very top of the cerebrum is for perceiving information from the body itself such as perceiving touch, heat, pain and so forth. Then there is a small lobe just over the eyes where we experience taste and smell.

These four lobes of the cerebrum do not "think" at all in the way we understand thinking, logic, and reason. They all feed their information to the frontal lobe of the brain. The frontal lobe is not associated with any particular sense, but it is also fed from centers found in the midbrain that encompass emotion and primitive responses. The frontal lobe combines all this information into what we call abstract thought.

Abstract thought is "thought" about anything we cannot hold in our hand or perceive through our senses. Thus it incorporates logic, reason, and all ideas that are abstract. That does not mean that abstract thought is always logical or reasonable as evidenced by human behavior.

The bottom line is that logic and reason happen, if they happen at all, in one of five lobes, in one of four layers, of the brain. The influence of the previous layers and other lobes is only slightly understood at this time. It would be reckless for either theologians or scientists to make large claims about those influences.

Yet, it seems to me that a number of people continue to make unreasonable, illogical, and concentrated attacks on one subject or the other. Why?

Chapter 19 – The Bad Guys

I don't know . . .

Why there IS a war between religion and science. I have argued in this book that there need not be one. As with many wars, there is an aggressor. In this war, it is clear who that is. I hope it has been shown that while atrocities may have been committed on both sides, and there exists bad blood, the true initiator of the conflict is neither religion nor science. No, there is a third, terrorist-like group attempting to manipulate the global stage to their advantage.

These are the militant, atheistic terrorists!

First the Innocent

As I began this book, I told you about John William Draper and his extreme, poorly-written attack on religion in 1874. His book was influential at the time and spawned the concept of a war between the two groups of people. The arguments and conclusions drawn between scientists and theologians have been legitimate topics for discussion, and fascinating contrasts have been made in the ensuing century and a half. But it has scarcely been a war.

I do not fault science as the aggressor because most scientists are, well, scientists. They try their best to think clearly, acknowledge their basic assumptions, consider the data, and formulate other hypotheses. This leads them to understand most of the clarifications I have made in this book, and they attempt to avoid the hubris that afflicts humanity. Mr.

Draper was not the first person to argue vehemently against religion. He was merely one of the first to utilize science as a cover for what has obviously become militant atheism.

Religion, likewise, cannot be considered the aggressor. While many religions proselyte and attempt to convert others to their way of thinking, very few that I know of have out-rightly rejected science. While some scientists may reject religion, religious people seldom forgo the use of electricity or any other benefit stemming from science. Most religious people see science as a window into a better understanding of God and their relationship to him. They at least accept science's beneficial effects on mankind.

The Guilty

However, in recent years, some atheists have become militant to the extreme. A long string of authors and thinkers have been public and combative. They have abandoned the reasonable language of science and fought with personal attacks, inflammatory language, faulty arguments, misleading assumptions, and downright lies.

I should point out that being an atheist does not make one anti-religious or even a "bad" person. Many atheists peacefully go about their own lives, often doing much good and behaving as morally and ethically as any religious person. This is not an attack on atheists. <u>It is a counter attack against militant atheism.</u>

These academic and spiritual terrorists almost universally cloak themselves in the garb of science while simultaneously ignoring scientific reason, language, and method. Many who claim scientific atheism are not even scientists. Others are scientists who have totally abandoned reason and objectivity though they may be sincere advocates for what they believe is true.

The motives of these public and militant atheists are interesting, and the they usually announce their purposes themselves. Their true objections are seldom scientific. In summary, most common arguments against religion, or against certain scientific theories, are not truly logical or even supported by clear reason. These arguments certainly are not couched in humility and do not recognize mankind's limited ability to know.

Atheists come in many different forms, and they have their own sectarian camps. They go by such names as secularists, non-believers, non-theists, agnostics, skeptics, free-thinkers, and humanists. There are fine points of argument that separate these groups. But for the most part, they are de facto atheists in one way or another. Of course, there are also atheists who live without any deep thought of scientific matters.

The remarkable traits shared by many in these militantly atheistic groups in today's world are intellectual-militancy and moral-self-confidence. It is easy to observe their arrogance although they, themselves, seem to be unaware of how they are perceived by others. It is simply this. They claim that every tenant of every religion concerning deity is wrong. They are the only ones who possess the truth.

No true scientist can make such claims since it is not within the power of science to answer theological questions. No theologian, or any Christian believer, rejects every other religion's claim to spiritual enlightenment or understanding of morality. Christians often bicker about details of doctrine, but most recognize that all faiths have uplifting beliefs and practices, even non-Christian faiths. Only the atheist claims omnipotence.

For several years I have written a science column for a local paper. Whenever I have made reference to my personal religious beliefs ever so slightly, I am inundated with responses correcting me. They suggest I am not free to say such things and even attack me personally. They know best, and I had better shut up.

However, not all terrorists wear masks. Prominent atheists have published a string of books in recent years filled with advocacy, not science: Richard Dawkins, The God Delusion (Dawkins, The God Delusion, 2008); Sam Harris, The End of Faith (Harris, 2005); Victor Stenger, God: the failed hypothesis (Stenger, 2008); Christopher Hitchens, God Is Not Great (Hitchens, 2009); and Michael Onfray, Atheist Manifesto (Onfray, 2005). These are joined by other writers who have anti-religion, especially anti-Christianity, agendas. A little less blatant in their pursuit are E. O. Wilson, Carl Sagan, Daniel Dennett, and Steven Pinker.

Throughout the writings of these men is a deep disdain for the common man and especially for those with religious beliefs. They, themselves, wish to be called "Brights" rather than atheists. The implication is clear. They are smarter than you and me.

They are not subtle. Let me give you some examples of their "reasoned", "thoughtful", "scientific", and "humble" opinions.

- Steven Weinberg wrote 'Anything that we scientists can do to weaken the hold of religion should be done and may in the end be our greatest contribution to civilization". (This is obviously a man who already believes he has contributed greatly to civilization.)
- Sam Harris wrote, ". . . the lunatic influence of religious belief." (Measured, reasonable, analysis that . . .)
- Christopher Hitchens wrote, "All religions and all churches are equally demented in their belief in divine intervention, or even the existence of the divine in the first place." (He writes with absolute certainty, beyond scientific analysis.)
- Dawkins wrote," The great unmentionable evil at the center of our culture is monotheism." (He disregards nuclear holocaust and other scientific maladies, of course.)

So while they claim science and reason as their front, their words and behavior belie any pretext to careful and thoughtful analysis. These quotes sound more like the rantings of a demagogue using language to whip up the crowd into a Nazi-like frenzy. These are not isolated quotes taken out of context. Their writings and books are filled with such unscientific statements and conclusions.

The danger here is that many people do not understand the limits of science and are influenced by these charlatans. Such people are convinced by arguments that are unreasonable from a scientific point of view.

We need to look at one other tactic atheist's use. They pronounce theological beliefs, as if they understand them, and then ridicule those beliefs. They ridicule beliefs that they, themselves, do not hold to or understand. If you want to know what any group believes, it is best to ask the believer, not the adversary. Consequently, atheists greatly misapply the teachings and conclusions of religion.

Often they state religious beliefs in terms of a ten-year-old child's understanding and then ridicule that. When a person tries to explain true theological principles, they claim it is too complicated and difficult. Like scientific arguments are simple?

These practices are not accidents or mistakes. They are strategies, intentionally utilized to persuade and convert, as surely as religions attempt to do the same. In fact, atheists have an almost universal set of principles, or atheistic catechism, to which they prescribe with at least as much fervor as any religious zealot.

- There is no God. (Only God knows.)
- There is no afterlife. (As if they have been there.)
- There is no way of knowing anything except through reason. (They know this without reason?)
- Nothing is more important than facts. (Except, of course, their opinions.)
- Explanations are proofs. (Only their explanations are to be considered.)

- Humans are nothing more than collections of chemical reactions. (Their chemical reactions are infallible. Yours aren't.)
- There is no such thing as free will. (They had no choice as to whether or not they should take this stand.)
- Religion is the source of human inhumanity. (Disregarding all secular atrocities. . .)

I hope to have shown how these beliefs are not defensible, scientific, or reasonable. In many cases, their conclusions are not even sensible. Next I will try to explain who these people are, why they believe as they do, and why they are so militantly disposed.

Chapter 20 – Why Does Atheism Exist?

I don't know . . .

I don't know "how" or "why" people adopt their individual beliefs. I know even less about why whole collections of people adopt a set of beliefs. Why do Catholics believe as they do? Jews? Hindu's? I suppose there are as many reasons for an ideology as there are people who profess it. It seems dangerous, illogical, and unreasonable for me to assign motives and understandings to other people.

However, not everyone suffers from this same reticence. Atheists seem to know exactly what motivates religious believers.

Atheists Explain Religion

Almost all militant, atheist writers have no trouble explaining religion. They believe religion originated in the ignorant past when humans were too dumb to understand how the world worked. They explain that ignorant humans created superstitions and elaborate rituals to appease what seemed to be a senseless and unpredictable world.

Atheists claim that these beliefs were perpetuated over the centuries by others who found sources of power and influence through the practice of religion. In other words, religious leaders are all charlatans; and those who profess religious beliefs are dumb, frightened, and easily duped.

They believe that because science was successful in explaining, or at least telling stories, about the workings of nature, that religion was shown to be false. Because they are such firm devotees of evolution, they think the purpose of life is to perpetuate genes.

They further believe that they represent the pinnacle of these advancements, and they now know enough to declare the meaning and purpose of life and of all things. The rest of us are quite simply "not too smart" or are "throw backs" to an earlier more ignorant time. After all, progress is only meaningful if it means advancement. They further believe that science, secularism, and the *intelligencia* have had nothing to do with creating all the suffering and atrocities of our present modern world.

They believe that, on the face of it, religion seems useless from an evolutionary perspective. Religion requires time and money to benefit a community, and often the benefit is for total strangers. Religious people give money, and sacrifice effort, to build cathedrals and burial sites. They don't work on some holidays, and they practice numerous other behaviors that seem to be a drag on reproduction.

Religious tenants provide a serious dilemma for atheists. How can they explain the continued existence, even expansion, of religious thought and believers of all kinds? These practices are not explained by their theoretical framework adopted from Darwin. Further, they hope that, if they can explain the roots of religion, they could undermine its authority. I am not guessing that this is their motivation. They have stated boldly that their goal is to abolish religion.

Their explanations beyond historical supposition of ignorance and superstition are poor at best. Dawkins has actually postulated that religion might be caused by "hyperactivity in a particular node of the brain". Others treat religion as "wishful thinking". Of course, no scientist has found that particular part of the brain or defined wishful thinking.

A similar proposal was put forward by Steven Pinker. He proposes that there is a "God module" in the brain that predisposes one to believe in God. He suggests that it serves no useful purpose but developed as a byproduct of some other trait that was necessary for survival. Of course, there has been absolutely no evidence of such a "module".

Evidence never seems to bother atheists as they propose such far-fetched hypotheses. Might it not be just as likely that there is a "Darwin module" in the brain that presupposes one to unbelief? This Darwinian Module might simply be an unfortunate accident of evolution, survival, and progress.

The Evolution of Religion

It is informative that the practice of religion has never completely disappeared from any society. It could not even be eradicated in totalitarian Russia though that country did its best to destroy it. This continuation of belief is a great surprise and presents a common dilemma for atheists.

It is not necessary to explain the origin of religion in scientific terms. Religious belief never stands or falls on logic and reason, but it inevitably involves questions of faith and free will. There is, however, an economic basis of religion that provides some answers for the perpetuation of religious belief, if not for the various forms of belief. I offer this explanation not as a necessary justification for religion. God is the justification for religion. But I offer this theory as a solution to set the minds of my atheistic friends at ease.

The Reverend Randy Alcorn, founder of the Eternal Perspective Ministries in Oregon, has suggested two, alternative, creation stories. He then rhetorically asks if it matters to the survival of the fit which is true. (Based on the account in Dinesh D'Souza (D'Souza, 2007).

Imagine you are descended from a tiny primordial cell and have evolved by chance and natural forces over billions of years into a grab bag of chemicals. You are a tiny being, on a tiny planet, in a tiny solar system, in a meaningless universe. You have no essence beyond your body; and when you die, it is the end. You came from nothing, and you are going nowhere.

The other scenario is that you are a special creation of a caring deity. You have the capacity to think, feel, worship and choose. You are unique from all other animals, and you are even unique among your kind. You believe every act you do is consequential to your god and your community. You were created with a purpose and are going towards a destination.

It would seem that the first scenario would produce people who do not know why they exist, how they think, what they are to do, or how they are to do it. The second group of people would be purposeful, committed, and goal oriented with an animating sense of purpose. Which community survives?

Wouldn't it seem like the first group might produce a listless people with no uniting or driving goal other than to, perhaps, rail against the other group? Such a group would have no great motive to produce children, the supposed purpose of atheistic life. If children are produced, there should not be so many as to interfere with the enjoyment of the immediate day-to-day living of life which has no other purpose aside from enjoyment and pleasure.

As a side note, it is obvious that, as birth control has become feasible and available to large numbers of people, secularists and atheists promote birth control. They often use the quasi-scientific idea of "overpopulation" as a rationale.

Of course, they overlook the fact that overpopulation is a result of longer life spans and not increasing birth rates. The truth is, decreasing birth rates are a direct threat to the production of the minds and people who create ideas, inventions, theories, and services that promote the welfare of humanity.

Atheists think it is the poor people who have large families, yet others in these societies are decidedly poor. Look at Russia, Europe, China, and other secular societies. These societies are decidedly poor with atheistic governments that promote birth control and forced abortions. When examined closely, it seems that religious communities like the Muslims, Mormons, Catholics and others, who have the large families, are the ones who seem to make evolutionary sense. Perhaps religion doesn't need an evolutionary explanation. Perhaps atheism does.

The Evolution of Atheism

There are profound consequences to insisting that man is nothing more than "matter" operating according to physical laws. One of the largest challenges is in what to believe about free-will. Most atheists conclude that free will is an illusion.

Richard Dawkins has said that mankind is nothing more than the method our genes use to create new genes. However, he also goes on to say, "We have the power to turn on our creators. We, alone on earth, can rebel against the tyranny of the selfish replicators. . . understand what our own selfish genes are up to . . . we may then at least have a chance to upset their designs." Does he mean that man has no free will, or that he has the free-will to decide to have free-will? This is nonsense.

Dawkins has no explanation for why man occupies such a unique place among animals. He does not even seem to think that such an explanation is needed. He may not believe that man is a creation of God; but he certainly acknowledges man as the most unique living creature, with no explanation of how this came to be.

Atheism appears, to me, to be similar to homosexuality in the scientific and evolutionary scheme of things. Why would nature select people who mate with others of the same sex? There is no selective advantage to the individual, the species, or the community in perpetuating homosexuality. In this case, it isn't a question of morality, just uselessness.

As perplexing is the question of why nature might select for a group of people who see no higher purpose in life than the universe, science, chemicals, or fleeting experiences that will end with no memories of existence. It would seem that such a dismal belief would inhibit reproduction as well as the willingness to invest in the next generation. In fact, this seems to be the case with many, if not most, atheists.

I don't really intend to refute atheism. I do want to show that there are alternative ways of looking at these questions, besides the way they are often portrayed, especially by those with a stake in the argument. Militant atheists have tried to move the argument in one direction and, to a great extent, have been able to do so. Obviously, they cannot withstand the scrutiny that using their own method brings.

Why do Unbelievers Not Believe?

Atheists constantly proclaim that their positions are based on reason and logic. If you listen to them, or read their books, they would have you believe that their sole reasons for rejecting God are reason and logic.

However, I believe I have shown that atheism is, by far, NOT the only reasonable alternative to belief. In fact, many of the conclusions and theories put forward by atheists are not reasonable or logical and are far from the only possible explanations available. Their own reasoning and arguments are often faulty and filled with the very emotional language and value judgments that they abhor in religious discussions.

In addition, non-belief is a dismal ideology. Strangely, many atheists embrace this aspect of atheism and portray themselves as lone, heroic figures courageously facing the cold, infinite night of nothingness. I think they see themselves as a kind of "Marlboro Man". For the younger audience, the Marlboro Man was the guy in a cigarette advertisement. He was a rugged cowboy, sitting on his horse at sunset, bravely smoking his cigarette while surveying the land just before dark. I've always thought the pose was a little comic, and the analogy is a little too accurate. He's dead now.

Darwinism is partly opposed by some on the grounds that it is a morally bankrupt theory. Do we want to imagine that we are animals who spend our time conniving and conspiring to perpetuate our own genes into future generations over those of our rivals? Steven Jay Gould has declared that, "We may yearn for higher answers – but none exist." In other words, there is no morality. He knows this for a fact and can state the future with absolute certainty. Even theologians argue over these things.

Gould continues, "This explanation (atheism), though superficially troubling if not terrifying, is ultimately liberating and exhilarating." I wonder what we are liberated from. Many atheists claim they want to free themselves from the restrictions of religion, so they can practice virtue. That is certainly an admirable motive, but how do you evaluate virtue?

Then the question to be asked is "Why do you have to do away with religion to practice virtue?" One can easily practice virtue alongside religion. Many, if not most, religious people are virtuous. As a matter of fact, virtue, charity, and human kinship are all part of the Christian religion as well as most other theologies. One can be virtuous in the name of the brotherhood of mankind.

Perhaps it is time to examine the atheists' motives more carefully. They certainly see themselves as intelligent and noble souls. Their language and arguments portray them, far more frequently, as emotional and unreasonable. I would like to consider some of their motivations for disbelief. I do not lump all atheists together as they frequently do religious believers. However, there may be some unifying principles in their lives as well as explanations.

Atheists Motives
Let's look at some of the things that atheists have said about why they have adopted their belief system.

Men are unfair, so there is no God
I spoke to a student today who assured me that he was an atheist. When asked why, he explained that he had been raised Catholic and had been sent to Catholic schools where he had felt oppressed. He was even told how to vote in elections. If he didn't comply, he would go to Hell. He had not seemed to consider the possibility that God might exist, but his parents and school teachers were mistaken. He did not consider that the behavior of the adults in his young and limited world might have been in error, or that the poor behavior of people really has nothing whatsoever to do with the existence of God.

Bad things are God's fault
Darwin does not claim to have lost his faith because of natural selection. In his own writings, he

219

explains that he became an unbeliever because he could not tolerate the notion of eternal damnation. He made this statement shortly after the death of his ten-year-old daughter. The Church of England at the time taught that she would go to Hell because she had not been baptized. Like many scientists who have not thought deeply about religious subjects, he assumed her death was God's fault. But he also assumed that the Church was right, and his daughter was condemned to damnation. It does not appear that he ever thought to question church doctrine before he questioned God.

In no other human capacity do we blame an event like death on those who did not prevent it instead of the thing that caused it. It is entirely illogical. Yet many intelligent people continue to blame God for the experiences in their lives. Perhaps God didn't kill your child, bacteria did. A drunk driver killed your spouse, not God. Atheism, for these people, appears to be a form of revenge as they display great anger toward God.

Theirs is an old and ancient argument. Why doesn't God stop bad things from happening if he is truly both kind and omnipotent? This truly seems like a contradiction to many. It is interesting that many scientists like to use this argument when they lack understanding of religion, God, and most theological subjects. Because they start their thinking and arguments assuming there is no God, it is a little hard to take their arguments seriously about how they think He should behave.

For example, scientists and atheists are very concerned with "when", "where", and "how", the universe was created. They never ask "why" the universe exists. This is the perfect illustration of how initial assumptions influence logic and reason. If, from the outset, they assume there is no purpose in the universe, they never ask what the purpose of the universe might be. Occasionally, if an atheist is asked this question, they will reply that the question is not in the domain of science. If they think that the purpose of life is not in the domain of science, why do they think that the existence of God is in the domain of science?

There are alternative explanations for why bad things happen to good people. It is beyond the scope of this book to enlarge on them. However, as I have already shown, it is seldom the case that there are only two possible explanations. Atheists would have us believe that, because they do not understand God's purposes, there can be no God. I have shown that it is never that simple.

Yet many atheists have not experienced loss, pain, or any type of suffering. Indeed, I would guess the majority of western-culture atheists have lived in such a way as to minimize all of those things. Many are educated, free, and comfortably well-off. In fact, generally those who live amongst us, and who suffer most, appear to be more believing. There must be other reasons for rejecting belief than pain and suffering.

Religion as slavery

Almost universally, through cultures and time, atheists have spoken of throwing off the yoke of religion as if it was something that was suppressive and oppressive. It is obvious that, for a great many people, religion is seen as neither. In fact, religion is seen by millions as liberating and exhilarating. So we must dig deeper to discover what some atheists find so oppressive about religion.

I find the idea of "religion as oppressive" strange. In the first place, I am free to do as I please whether a member of a Church or not. Of course, I may not be free to participate fully if I choose certain actions, but that is a matter of consequences. I can choose my behavior, but I cannot choose the consequences of my behavior. If I defy gravity, it is going to hurt.

What does the Church demand of us that is so stifling? If there is no God, does it still make sense to lie, cheat, or steal? What does religion require of adherents that is so tyrannical?

The appeal of Darwinism and atheism, for many people, is that it frees them from the responsibility of a "higher" human nature or law and places them on a spectrum alongside other animals. As such, they are free to behave in ways that are animal-like without worry of consequence beyond the disapproval of some other animals. Animals are not held accountable for their actions in the same way that people are. It would be nonsense to speak of a "bad coyote". Coyotes are just coyotes. I suppose, if a coyote did not function as a coyote very well, it would die. Maybe we could say of a dead coyote that it was a bad coyote.

A basic Christian tenet is that death does not bring annihilation, but accountability. We will be judged as to whether or not we were "good" or "bad".

In fact, in Genesis God proclaims all of his creation "good", except humans. I think He does this because it remains to be seen whether we will be good or bad.

If one wants to escape this threat of being judged poorly and of not being held accountable for actions taken during his life, the very best recourse would be to do away with the judge. Atheists often claim that religion is the "opiate of the people". Nice phrase, but it's meaningless.

Ignoring the judge and pretending that there is no accountability is truly like taking drugs, making atheism the true opiate of the people. Even if one does not believe in life after death, most people believe in accountability. Without religion, there remains no higher law and, therefore, no accountability.

I do not mean to imply that atheists are not good people. However, if you want to live a degenerate life, God is your enemy! Atheists may not be more evil, but atheism provides a nice hiding place from which you do not have to acknowledge lies, murders, theft, adultery, greed, gluttony, selfishness, covetousness, idleness, or other deadly sins.

Nietzsche was perhaps the most militant atheist of all. Modern atheists like to claim that the death of God does not mean the death of morality. However, that is exactly what Nietzsche believed! He maintained that any goal was legitimate, and the effect it had on others was irrelevant. Our only purpose was to pursue our lives with all our energy and determination just like other animals pursue theirs.

You may think I am supposing too much and attributing motivations to atheists that they do not possess. Aldous Huxley, atheistic grandson of Thomas Huxley and Darwin's champion, said, "The liberation we desired was . . . liberation from a certain system of morality. We objected to the morality because it interfered with our sexual freedom." (Huxley, 1996)

Bertrand Russell wrote, "The worst feature of the Christian religion is its attitude toward sex." (Russell B., 1957) Christopher Hitchens wrote, "The divorce between the sexual life and fear . . . can now at last be attempted on the sole condition that we banish all religions from the discourse." (Hitchens, 2009) Steven Pinker, has said of abortion, "a capacity for neonaticide is built into the biological design of our parental emotions." (Pinker, 1997) How he determined this to be scientifically true makes for an interesting question.

If there is no judge, there is no judgment. One is left to believe, from these and many other statements from prominent atheists, that theirs is not an intellectual revolt. It appears to be a revolt against morality.

There is one final argument about the tyranny of religion. I belong to a religion that many people view as being very strict. Peoples' primary reason for this belief is that we abstain from alcohol, tea, coffee, tobacco, and sexual relations outside of the bonds of marriage. I realize there are other religions that adhere to these same restraints, but I can only speak to my own experience. I am always perplexed when I see, or hear of, someone who leaves our particular faith and is now smoking or drinking.

I can fully understand someone who has decided they do not believe the tenets of a certain belief system. They should leave that system. What I don't understand is how a person proclaiming different tenets suddenly declares it wise to smoke tobacco or drink excessive alcohol.

I don't understand how indiscriminate sexual activity makes a person any happier and his/her life any richer. I see a multitude of ways in which all of these practices impoverish and complicate a person's life. Yet atheists never seem to address the issue of how being free from moral restraints makes the average person's life richer, happier, longer, or more peaceful. They also fail to explain how such deviations make communities and cultures stronger and more prosperous.

Atheists are free to not believe, of course. But I am not sure why they think that living a Christian life is in in some way bad for others. Of course, many atheists live Christian standards of morality, but the militant atheists seem to believe that people will be happier and better off if they do not adopt any strict standards of behavior. Yet, immorality does not make good sense.

Chapter 21 Atheism and Morality

I don't know . . .

I don't know why I try to "be good". Do I try to "be good" because of the moral law given by God and the manmade law that threatens punishment? Or do I try to "be good" because I really don't want to hurt anyone if I can help it? I have hurt peoples' feelings. I have hurt people physically. I have run red lights. I have told lies, and lying is against God's law.

However, for the most part, I don't go around trying to do those things. If I purposefully run a red light, it is because the light didn't trip for three minutes at 2:00 AM in the morning. I looked both ways carefully. Of course, I don't know if I was checking to be safe, or if I was checking to see if there was a policeman around.

Theologians tend to think they know pretty well what morality is, but they are generous in overlooking it. Atheists are pretty sure they can be moral without religion, but they generally don't agree about what morality is.

What is morality?

Religion and morality are not necessarily the same thing although religion is often a champion of specific moralities. Many atheists claim that they can be, and are, moral people. In fact, they often claim to be more moral than religious people. Atheists often fault religious people as hypocritical for failing to live up to the strict codes they espouse. They assure us that morality is quite possible without the need to follow any edicts from God.

226

The problem is that they are both right and wrong. There are wicked people who profess religion. However, there are also wicked people who profess no religion. Those who site the Crusades or the Inquisition usually ignore Mao Tse Tung, Stalin, and Hitler. The differences in the death tolls of the two groups are staggering.

Likewise, there are good Christian men and women, AND atheists, who are decent and moral. The atheists claim morality is manmade, created from individual experiences and community attitudes, not a set of behaviors dictated from above.

The first question becomes, "Is there a universal or objective morality?" Humans seem to have become accustomed to thinking of morality as subjective, a statement of personal opinion. Generally, we consider objective information good but subjective information suspect. This despite the fact that when we really need to know what to do in our lives, we usually rely on the advice of trusted family and friends, subjective data, far more than scientific, objective data.

However, what if there is an objective morality? In fact, it appears that there may well be. No other animal besides mankind appears to have a moral consciousness. There is no evidence that even our closest animal relatives weigh their own interests against the rights of others, or that they develop some concept of the greater good for society. Yet all human cultures have some things about which they might say, "You shouldn't do that." or "Bad boy!"

Another large difference between humans and all other animals is that non-humans obey scientific laws. If one offers a scrap of meat to a well-fed dog, the animal will most likely wolf it down. Throw a stone into the water, and it will sink. There is no choice to be made in these matters.

There are parts of humans that operate according to physical laws. If we jump from a third-floor window, we will fall. It would make no sense to tell someone they "shouldn't" fall. However, we would undoubtedly say they "shouldn't" jump. Words like "should" and "ought" imply some kind of shared standard. Laws of nature cannot be broken. Laws of morality may be violated.

It would make no sense to say, "Thou shalt not kill" if there was no option. When we say that one shouldn't lie, we are saying that there is a possibility than one could. This only makes sense, however, if there is a shared judgment about the importance of these things. You and I both agree that killing is not what we should be doing.

Atheists like to claim there is great diversity in moral agreement in our own culture and between cultures. They claim there is no universal, objective morality. These claims are based on faulty thinking and faulty facts.

Atheists like to portray themselves as independent, reasonable, and logical. Yet, suppose there is a society where the Ten Commandments are universally and systematically broken. Is that, in itself, proof that there is no universal morality as they claim? Of course not! At best, it proves that this particular culture is different from our culture. At worst, it proves that this culture is not a nice place where most of us would want to live, even atheists.

In fact, we live in a country where the Ten Commandments are systematically broken, pretty much universally ignored, and institutionally rejected. What we are left with is not the diversity of moral law but the fact that many people fail to practice that set of moral laws. As usual, there are multiple, logical ways of interpreting experience and data. I have tried to show that atheists refuse to acknowledge this multiplicity of interpretations but insist on straw men to bolster their arguments.

What if this hypothetical "foreign" culture didn't believe in honoring parents, and parent beating by adult children was routinely practiced? Would that disprove a moral law? In case you are not sure, imagine if that same culture disagreed with the heliocentric theory of the planets! Would that disprove the heliocentric theory? Peoples' actions and beliefs would not show that the law was wrong, only that the people were wrong. So while this argument by atheists sound plausible, in fact it is trivial.

On to the facts. No such amoral culture or society has been found. In fact, scholars have found that every society has a set of moral laws. Morality, it appears, is universal. While there are nuances between, and even within societies, every culture has a description separating what is from what ought to be.

Secondly, the moral diversity that the atheists like to site is vastly over exaggerated and perhaps doesn't exist at all. Every culture appears to adhere to some form of the golden rule, "Do unto others as you would have them do to you." This universal set of standards, shared by almost all cultures, has existed throughout time as well. The Romans and Greeks ascribed to many of the same moral beliefs as modern-day Jews, Christians, Muslims, Hindus, and Buddhists.

There is great diversity in moral practices but great uniformity in moral beliefs. There are societies where lying and cowardice may be excused, but they are never upheld as honorable. In one society, a man may have only one wife; but in another, he may have three. Yet all groups agree on family restrictions and moral obligations to provide for families and children.

Relative Morality

Atheists claim all morality is relative. As such, morality then becomes relative to the individual and what is in a heart. Each individual is allowed to live his or her own moral principles, but they're not obliged to live any moral principles handed down from above. This line of thought is dangerous because it is based on poor assumptions.

Reason begins with assumptions, and this belief begins with several. First, is the assumption that there is no universal moral law. I have already shown that to be, at least, doubtful. This belief also assumes that variability of moral beliefs proves there are none. I have already shown that is not a reasonable conclusion.

Since relative morality is determined by the individual, it denies any obligation for the well-being of others. What makes the individual happy, or what is good for others from the individuals' perspective, is what should and ought to be taken into account. This is a complete reversal from what most of humanity has thought morality to be: what should happen for the greater good. When the "I" is the creator of what should be, the greater good is not considered.

Those who think that morality should arise from within themselves begin to think and use language in a peculiar way. If they determine what should and should not be, then what they determine is what should be for all others regardless of what the others think. At best, this leads to conflict between individual moralities. At worst, it leads to tyranny.

However, relative morality also assumes that humans are all "relatively" good, and their desires and moral decisions will always be good for others as well as themselves. No matter how much I would like to think this is so, cold hard data suggests that this is not a universal truth.

The fact is, if relative morality is all that there is, we need a new term for what has traditionally been called morality. If any man declares what is right or wrong, I can always counter with the argument that he is not in charge of me. This used to be a common retort of children. "Oh yeah!? Well, you're not the boss of me!"

Morality can only exist as an absolute concept if it is proclaimed by one with authority over all men. That would be God. Then there can be no argument. Behavior is either moral or not. Anything proclaimed by man is immoral or needs another name.

Religious Morality

Atheists have been very successful at convincing the world that religion has been the major cause of pain and suffering in the world. The following are quotes from a variety of atheists. Sam Harris has called religion, "the most potent source of human conflict, past and present." Steven Pinker said that "religions have given us stoning's, witch-burnings, crusades, inquisitions, jihads, fatwas, suicide bombers, and abortion clinic gunmen." Interestingly, he doesn't include Communism, Stalinism, Maoism, or abortion clinics!

Robert Kuttner wrote, "The Crusades slaughtered millions in the name of Jesus. The inquisition brought the torture and murder of millions more." (Kuttner, 2004) Richard Dawkins contends that, "most, if not all, of the violent enmities in the world today are due to the divisive forces of religion." (Dawkins, The God Delusion, 2008)

The first problem, of the usual litany of claims of religious violence, is that the claim is highly exaggerated. Atheists often place blame on Christianity when other forces were at play. The Crusades, for example, were religious in nature. However, the Crusades occurred after three centuries of Islamic expansion and the invasion of Europe by Mediterranean countries.

Each situation was far more political than religious. In truth, prior to the rise of the Islamic Caliphate, the Middle East was predominantly Christian. The Islamic Caliphate invasion and resultant crusades were far more about political power than religious belief. Religion? Yes.

Christianity? Well, if self-preservation is wrong, then yes.

The inquisition was not exactly what we in the west have been told it was by atheists. It is beyond the confines of this book to address the entire truth of the inquisition, but here are some salient facts.

The Inquisition did not target Jews unless the Jews had converted to Christianity. The inquisition was actually a fairly-benign tribunal. By far, the majority of penalties handed out were in the form of penance such as fasting or community service. Only about 2000 deaths for heresy actually occurred, and those deaths took place over a period of350 years. That's about five deaths per year. I believe those deaths were all wrong. The point, though, is that the inquisition was not the wholesale slaughter that it is usually purported to be by atheists.

Carl Sagan wrote in his book, *The Demon-haunted World*, that "perhaps hundreds of thousands, perhaps millions" of people were burned as witches. (Sagan, The Demon Haunted World, 1996) However, we don't know what his source for these numbers were. He didn't site anyone. Is that scientific?

Historians and non-scientists, who have studied this matter carefully, put the number of deaths at one hundred thousand, or less. While that number is terribly high, those deaths occurred over a period of time of thousands of years. In addition, the suspicion of witchcraft is held in all societies, and persecution of witches has nothing to do with religious belief, only with supernatural belief.

Sadly, it is a similar number that were killed by each atomic bomb at Hiroshima and Nagasaki.

Neither Christians, nor believers of any other religion, were responsible for that event.

Atheism and Human Suffering

I could go on, but I think it would be more helpful to talk about the role atheism and relative morality have played in human suffering. The Inquisition occurred over 500 years ago, and it is still an issue. But atheists never seem to mention Chairman Mao, Stalin, or numerous other Godless tyrants of the last fifty years. Many atheists have attempted to blame the atrocities committed by Hitler on religion. Any well-read person knows Hitler was not religious, and he simply used the German church as a tool for gaining control.

It is estimated Stalin killed twenty million people. Mao Zedong is reputed to have killed seventy million people. Even if these numbers are over-estimates, they are still completely staggering. Would it matter if the numbers were off by half? Not when you consider that these deaths were of the two men's own peoples. These deaths did not occur in war, but in peacetime, to solidify authority. Hitler was a piker, in comparison, with only about ten million deaths attributed to his regime.

These numbers do not even include the assassinations and murders committed by numerous lesser tyrants such as Lenin, Khrushchev, Brezhnev, Pol Pot, Nicolae Ceausescu, Castro, or Kim Jong-il. The "big three" atheistic regimes, alone, killed over one hundred million people! There are, of course, more people to kill and increasingly sophisticated methods of killing in modern times. At the same time, it is estimated that the world population has increased by about five times from the times of the inquisition to today. The inquisition and the crusades, together, are estimated to have killed about two hundred thousand people. If we increase that number by five times to compensate for population growth, it would have been about one million deaths by today's numbers. The crusades and Inquisition occurred over a five-hundred-year period. So all religion-related deaths account for about one percent of the deaths caused by atheists in the last one hundred years alone.

It seems logical to ask, then, if there is something in atheism that causes people to do wicked things. After all, atheists seem to believe that there is something in religion then specifically leads humans to be wicked. Atheists like to claim that wicked acts done by atheists are not done in the name of atheism, but wicked acts done by the religious are done because of religion. It doesn't seem like one should have it both ways.

Steven Weinberg, a physicist, concedes that atheists have done their share of wicked deeds. But he argues that whenever science has been invoked to justify those deeds, it has been a perversion of science. It is interesting that non-believers do not concede that, perhaps, when wicked things are done in the name of religion, it is a perversion of religion.

Daniel Dennett wants religion judged by the consequences of religious belief, "by their fruits ye shall know them". (Dennett, 2007) He does not care if the consequences are enacted by people who practice the best tenets of the religion or if the consequences were intended by the teachings of the religion. If a person does an evil act and says he does it in the name of religion, then it is religions fault.

If we grant the logic of such an argument, would it not also be fair to claim that all one hundred million deaths of the past centuries caused by atheistic men must be the responsibility of all atheists? Again, atheists cannot have it both ways: Religion is responsible for the acts of the irreligious acting in its name, but atheists are not responsible for atheists' acts.

Of course, if "by your fruits ye shall know them" is to be followed, then all the good that has been accomplished in the name of religion must also be weighed and acknowledged. For example, while most atheists claim to be moral, I know of no atheist soup kitchens, hospitals, service organizations, or charities. In fact, it has been repeatedly shown that atheists donate far less to moral causes and the welfare of other humans than religious people do.

The question remains, however, whether religion or atheism is the cause of violence, murder, and wickedness. Many atheists want to claim there is nothing in atheism, itself, that would lead to wickedness. So how can the absence of belief cause social harm?

How Atheism Goes Wrong

Atheists seem to believe that because they do not believe in religion, they have no beliefs. Of course, that isn't true. They have a set of beliefs. They believe in science, reason, logic, and progress. Since their belief is that reason, logic, and science are superior in answering some questions and in accomplishing some tasks, they are superior in all things.

They are firmly convinced that their world view is the superior world view. Because their understandings of the world are superior, they must be right, and all others must be wrong. They almost sound religious! They are the ones who understand the grand, noble, inexorable, and incontrovertible forces that govern the universe.

Of course, progress implies a goal. If one does not know where they are going, it cannot be said they have made progress. The atheists are sure that their goals are what everyone should strive for, and they measure progress by their own vision. Even if they should arrive at the utopia they envision, wouldn't they just find that there are others there who see a different vision?

They are as positive in their outlook as any religion. So positive are some atheists, that they are correct that any act in behalf of their beliefs is excusable. Hitler simply applied the science of "the breeding of the fittest" as he saw it. Stalin simply knew he was right, and all opposition had to be crushed. The French Jacobins worshiped the goddess of reason, and they oversaw The French Revolution. In the view of Jacobins, and many atheists, "unreasonable" people can be put to death!

Their thoughts of superiority go even beyond this assumption. Since an atheist's morals are relative, there are no uniform moral restraints. Each man should be able to decide for himself what is acceptable and good for the community. Since the atheist belief system is superior to those of others, atheistic beliefs do not rely on outside authority. That means there is nothing to condemn any action on an atheist's part. Anything he sees as beneficial to the whole, no matter how much the whole or parts of it may disagree, must be enacted.

The atheistic prophet, Nietzsche, foretold the next two centuries with striking accuracy. He predicted, in the nineteenth century, that the next two centuries would see wars and violence beyond anything imagined at that time. He knew that the "death of God" would mean the end of shared values and moral restraint. How could he have foreseen this had he not seen the connection between atheism and wickedness?

Let us suppose for a moment that Nietzsche's concept of a superman was actually achieved. He envisioned the development of a "superman", superior to common man in every way; physically, mentally, and morally. Suppose that the government and medical science found a way to breed and teach a wonderful, new, powerful mankind. I would guess that the first thing any strong and powerful man or woman would do is throw off the government that restricted them.

Violence and carnage have been the fruits of atheism far more than the fruits of religion. If indeed, as Dennett and other atheists suggest, we judge religion by its consequences, then it is time to evaluate atheism in the same way. Either personal beliefs have nothing to do with communal good, or they do. If they do, atheism comes off the worst by comparison of factors of ten or more. Perhaps it is time to re-embrace religion as an essential step forward in establishing a more peaceful existence for humanity.

I hope it is clear that I have no animosity towards individual atheists. Each man may choose to believe as he wishes. But men do not have the right to hi-jack science and theology to excuse their own bad behavior, no matter how right they may think they are. People may choose to lie about history, to misrepresent numbers, and to reach conclusions in any way they see fit. Their doing so does not make it true or excusable to mislead and misapply reason and logic.

Part V - Epilogue

Chapter 22 - Truth

I don't know . . . what I don't know. It is a truth dating back to the Egyptians and the Old Testament that "Everything is made up of everything we know, and everything we don't know." Humans, at every age, think they know everything because they know everything they have ever known. They forget that there are things they have not known, and therefore, things they probably do not yet know.

Theologians may actually be even more susceptible to this flaw in thinking because so much of what they think they know can't be experienced in reality. Theologians often argue vehemently over differing points of doctrine that cannot actually be known. Often they exclude others over matters that are speculative or accepted on faith, while ignoring common fundamental beliefs denominations share. They will argue over the nature of the Godhead, or the nature of Heaven and Hell, while ignoring the fact that they all value moral behavior, charity, and strong families.

Scientists often fall into the trap of thinking they know more than they do. I have a friend who is a physician. He tells about the time he was summoned to help on an international flight because of a medical emergency on the plane. He is an oncologist who specializes in radiation treatment of specific types of cancer. He was absolutely terrified of dealing with an everyday-trauma event. He was well aware of what he didn't know in this respect.

Each new generation tends to think they are different. Mine did, and we were wrong. The present one does, and they are wrong as well. A twenty-year-old knows everything he has experienced. Since he has experienced nothing else, he cannot know the things he has not yet experienced. Yet he often thinks he knows. Just because we haven't experienced something doesn't mean that it does not exist.

There is also something called truth, and truth is not always ambivalent. There are scientific, material truths that are universal and reliable. Gravity, magnetism, life, the movement of the planets all seem to be true. Of course, we can learn to use these laws to accomplish much good or bad.

From common historical experience, it seems that there are also metaphysical truths that are pretty reliable. It seems to be true that lying, stealing, drug abuse, and violence all harm families as well as society. It seems to be true that certain social settings and practices such as worship services, prayer, meditation and serving others help us to be at peace and accomplish more good. Ignoring them can also cause much suffering.

Opposites

If there is truth, there is untruth. Because we may not discern the truth of a story, does not mean that there isn't one. There are absolutes in the world. There was a time before the Universe was created and also a time after it came into being. There may be order where we humans cannot perceive it. There is an up and a down. There is a north and a south. There is a black and a white, a dark and a light.

It seems obvious to both scientists and theologians that the Universe is not infinitely old. There was a time when there was no universe as we recognize it today. Then there was a time when there was one, and a chain of events began. If we were infinitely wise and knowledgeable, we could trace all events back to that beginning moment. We could call that moment "The Big Bang" or "Genesis". In either case, there is always a previous cause for everything.

There are things that are proclaimed to be truth that are not. At least, they are unreliable as landmarks. Much of science consists of proclamations, or tiny bits of facts, which may only be true if looked at from the specific perspective of science. Much of science is relative and pertains to only the physical manifestation of our experiences.

Let's take an example of a truth that is relative. The concept of left and right, whether used politically or directionally, depends entirely on which direction one is facing. If I am speaking on the phone and you tell me to turn left, I have no idea what that means unless you first tell me in which direction to begin. In fact, turning left could be exactly opposite of what either of us intend. It could be false. Left and right are not truths. They are relative truths.

Another example of unreliable truth is "kindness". Kindness sounds like something we should all strive for, except kindness is not uniformly defined. Just like left and right, kindness may depend on which direction we are facing or the circumstance we are in. Loaning money to a friend may be an ultimate statement of faith in his ability to repay. However, loaning money so that the friend can purchase harmful and illicit drugs is not a kindness. Loaning money reluctantly and with trepidation might feel somewhere in between. Either way, loaning money to friends surely changes the relationship between the friends.

Rich and poor are yet another example of "untruthful standards". Is one poor if they own an air-conditioned home and three big screen TV's? A significant number of "government defined poor" live like that. Is a person poor who chooses to work only twenty-five hours a week so they can ski? Should the person who lives without air conditioning, to save money for investment, be considered poor?

Using science and technology, we can measure the increase in carbon dioxide in the atmosphere. We cannot determine whether the carbon dioxide is there because of industrialization, environmental degradation, long-term cycles, reduction in photosynthesis, or atmospheric changes associated with causes outside of our atmosphere.

We can show colorful images of the brain while it is engaged in different activities. However, the brain is not the mind. Whether or not such images are predictive, prescriptive, or even interpreted correctly is almost impossible to conclude. Of course, the scientists engage in carbon dioxide research and neuroscience. Both have reason to claim their research is significant. If one is facing the right direction, I'm sure it is. However, like left and right, rich and poor, kind and cruel, scientific standards are often unreliable in determining truth and seldom helpful in living well.

The point is that there is "right" and "wrong", and the wrong is a negative concept. We need to be able to identify one from the other. Is it more important to recycle or to visit a friend in the hospital? Is it more important to encourage self-expression and personal choice or to encourage self-discipline and stable families and societies? Is it more important to reduce carbon dioxide emissions or to increase carbon dioxide consumption by planting more trees and growing crops more wisely? Is it more important to have courage than to have knowledge?

War Recap

My intent has not been to "prove" religion and "attack" science, or the other way around. I do not excuse religion of its excesses, but I also do not imbue science with omniscience. In our world, I do believe that religion needs to be looked at less skeptically, and I also believe science should be looked at more skeptically.

There has been little effort on either side to consider alternative explanations that fit the other discipline's paradigm. I hope I have enabled and encouraged the consideration of alternative positions and shed enlightenment on how both disciplines can improve the world.

The scientist's job is to answer questions about the physical world for which we don't yet have the answers. The Theologian's job is to answer questions about why there is evil, the nature of God, and the purpose of life. Sometimes there are answers, sometimes not, and sometimes there are simply no answers yet. Sometimes the answer to a question in one of the fields raises questions in another. None of that describes a war. Isn't it all fascinating!?

Works Cited

Aquinas, T. (1998). Selected Writings. New York, NY: Penguin.

Brecht, B. (2008). The Life of Galileo. New York, NY: Penguin Books.

Darwin, C. (2004). The Descent of Man. London: Penguin Classics.

Dawkins, R. (1995). River of Eden: A Darwinian view of life. New York, NY: Basic Books.

Dawkins, R. (2008). The God Delusion. Boston, USA: Martiner Books.

Dennett, D. (2007). Breaking the Spell. New York, NY: Penguin Books.

Dicke, R. H. (1965). Cosmic Black-Body Radiation. Princeton, NJ: Astrophysical Journal 142.

Dobzhansky, T. (2013). Mutation Driven Evolution. Oxford, Great Britain: Oxford University Press.

Draper, J. W. (1874). History of the Conflict Between Religion and Science: the classic examination of the ancient and perpetual collision. Richmond, VA: Republished by Creative Space Independent Publishing Platform.

D'Souza, D. (2007). What's So Great About Christianity. Carol Stream, Illinois: Tyndall House Publishers. Inc.

Elbert, J. (2000). Are Souls Real? Amherst, NY: Prometheus Books.

Eyring, Henry B. (1983). Reflections of a Scientist. Deseret Book Co; 1St Edition. Salt Lake City, Utah.

Feynman, R. (1998). The Meaning of it All: thoughts of a citizen scientist. New York, NY: Basic Books.

God. (2010). Genesis. Grand Rapids, MI: Zondervan.

Harris, S. (2005). The End of Faith. New York: W. W. Norton and Co.

Haskins, C. H. (1957). The Rise of Universities. Ithaca, NY: Cornell University Press.

Hawking, S. (1996). A Brief History of Time. New York, NY: Bantam Books.

Hitchens, C. (2009). God Is Not Great: how religion poisons everything. New York, NY: Hatchette Book Co.

Horgan, J. (1999). The Undiscovered Mind: how the human brain defies replication medication and explanation. New York, NY: Touchstone.

Huxley, A. (1996). Confessions of a Professed Atheists. Report.

Irving, W. (1828). A History of the Life and Voyages of Christopher Columbus, in The Works of Washington Irving. Anne Arbor, MI: University of Michigan.

Koestler, A. (1990). The Sleepwalkers: a history of man's changing vision of the universe. New York, NY: Penguin Books.

Kuttner, R. (2004, November). What Would Jefferson Do? American Prospect, p. 31.

Noss, D. (2011). A History of the Worlds Religions (13 ed.). Upper Saddle River, NJ: Pearson.

Onfray, M. (2005). Atheist Manifesto: the case against Christianity, Judaism and Islam. New York, NY: Arcade Publishing.

Paul. (1980). Hebrews. New York, NY: American Bible Society.

Pinker. (1997, November 2). Why They Kill Their Newborns. New York Times.

Roughgarden, J. (2006). Evolution and Christian Faith. Washington DC: Island Press.

Russell, B. (1957). Why I Am Not a Christian. New York, NY: Simon and Schuster.

Russell, J. B. (1991). Inventing the Flat Earth. Westport, CT: Praeger Publishing.

Sagan, C. (1996). The Demon Haunted World. New York, NY: Ballantine Books.

Sagan, C. (2013). Cosmos. New York, NY: Ballantine Books.

Schmidt, A. (2001). Under the influence: how Christianity transformed civilization. Grand Rapids, MI: Zondervan Publishing.

Smolen, L. (1997). The Life of the Cosmos. Oxford, United Kingdom: Oxford University Press.

Stenger, V. (2008). God: the failed hypothesis. Amhurst, NY: Prometheus Books.

Tarnas, R. (1993). The Passion of the Western Mind. New York, NY: Ballantine Books.

Whitehead, A. N. (1953). Science and the Modern World. New York, NY: Free Press.

Wilson, E. O. (1998). Consilience: the unity of knowledge. New York, NY: Knopf.

www.ingramcontent.com/pod-product-compliance
Lightning Source LLC
Chambersburg PA
CBHW051858170526
45168CB00001B/160